The Institute of Biology's
Studies in Biology no. 123

The Biochemistry of Pollution

J. H. Ottaway

D.Sc., Ph.D.
Senior Lecturer
Department of Biochemistry,
Edinburgh University Medical School

Edward Arnold

First published 1980
by Edward Arnold (Publishers) Limited
41 Bedford Square, London WC1B 3DQ

British Library Cataloguing in Publication Data

Ottaway, James Henry
  The biochemistry of pollution. – (Institute of
  Biology. Studies in biology; no. 123
  ISSN 0537-9024).
  1.   Pollution
  2.   Biological chemistry
  I.   Title   II.   Series
  574.1'92   TD177

ISBN 0-7131-2784-8

Printed and bound in Great Britain at
The Camelot Press Ltd, Southampton

General Preface to the Series

Because it is no longer possible for one textbook to cover the whole field of biology while remaining sufficiently up to date, the Institute of Biology has sponsored this series so that teachers and students can learn about significant developments. The enthusiastic acceptance of 'Studies in Biology' shows that the books are providing authoritative views of biological topics.

The features of the series include the attention given to methods, the selected list of books for further reading and, wherever possible, suggestions for practical work.

Readers' comments will be welcomed by the Education Officer of the Institute.

1980

Institute of Biology
41 Queen's Gate
London SW7 5HU

Preface

We have all become aware of environmental issues in the last two decades. Pollutants are chemicals; their persistence in or disappearance from the environment, and their biological consequences, can be explained fully only in biochemical terms. There are thousands of possible pollutants – the European Economic Community alone has a 'black list' of some 1500 chemicals – and it has been necessary to be very selective, or the book would have become a mere catalogue of unrelated reactions. Some of the more important xenobiotics are metabolized by pathways that have only recently been elucidated, and which are not fully described in most textbooks. Special attention has been given to these pathways. Otherwise it has been necessary, for reasons of space, to assume that the reader has access to a standard biochemical text.

A chapter on radiation hazards has been included, because this is a live environmental issue, and one on which it is difficult to find plain technical information. The book as a whole is designed to provide sufficient factual information at the biochemical level for the reader to make up his own mind on the issues involved. The author has tried to lay his own opinions strictly aside – with what success, the reader must judge.

Edinburgh, 1980   J. H. O.

Contents

1 Fresh-water Pollution

1.1 Introduction

Pollution of fresh water is one of the most serious environmental problems for the world as a whole. Unless circulation from the deep oceans is inadequate (see Chapter 2), heavy marine pollution does not extend more than a few miles offshore. The situation is very different for fresh waters, and it is difficult to see how many of the problems can, with present technology, be alleviated. We will restrict ourselves to surface waters, which may be divided into running streams, and lakes. Even in lighly-populated hard-rock areas, such as the Scottish Highlands, surface water always contains nutrients in sufficient quantities to support many kinds of life. The further a body of water (running or still) lies within a highly-populated area, the higher the concentration of nutrients that it will contain. It is not necessarily more active biologically.

Even in the absence of accidental pollution, there is not enough unused water in many temperate and tropical countries for the needs of the population. About 0.06% of the total fresh water in the world at any one time is in lakes and rivers – about 3×10^{15} litres. With a world population of 3000 million people, in theory everyone has a million litres to call his own. In practice, most of this water is not within reach of the major centres of population. In the U.S.A., the total stream flow is only 21000 litres day^{-1} per person, and the *actual* amount withdrawn from streams is less than 4000 litres day^{-1} per person. Even the larger volume is barely capable of dissolving enough oxygen to oxidize the nitrogenous compounds which the population of the U.S.A. excretes each day, leaving nothing over for the non-nitrogenous components, or the waste products of animals or of industry. As far as possible, then, these oxidations must not be allowed to occur in lakes and streams.

Industrial processes require vast quantities of water, often for cooling, but sometimes to dilute waste products to an acceptable concentration before returning them to rivers. The very large volumes sometimes quoted for industry's requirements must be assessed with caution – the water does not simply disappear, as ultimately most of it returns to source, and the water need not be drinkable. Nevertheless, effluent water from industrial plants is always warmer than the intake, and almost always more impure. The temperature differential is often the most important factor, as explained below. Power stations, particularly nuclear plants, are especially bad sources of heat pollution (see Chapter 5).

A serious source of industrial pollution is the leaching of sulphides (particularly pyrite, FeS_2) from mine workings or dumps. Dissolved O_2 converts this to $FeSO_4$ and then to ferric sulphate, which is hydrolysed.

Ferric hydroxide ($Fe(OH)_3$) is precipitated in the stream bed, and dilute H_2SO_4 flows off. The consequent high acidity and the loss of O_2 makes the water biologically sterile, so it is both useless as a diluent, and unfit to drink. In heavily polluted waters downstream the sulphate may be reduced by anaerobic bacteria to sulphide (see p. 12), which is both toxic and unpleasant. Salts, especially potassium chloride from salt mines, may also cause problems.

Up to a point, river water may be re-used. Estimates have shown that much of the water running from London's taps has been taken from the Thames, purified, and returned to the river at least four times previously. On each occasion the water that is returned, although technically unobjectionable, adds a quota of nutrients, particularly nitrate, so this re-use cannot go on indefinitely without dilution by 'virgin' water.

Dilution of polluted effluent by purer water is a simple thing to specify, but the purity characteristics of the diluent water need to be discussed in some detail. Water flowing from the drains or sewage works of a town can almost always become biologically and biochemically more pure, if time is allowed for microbial action to go to completion. This requires a large volume of water so that the concentration of pollutants remains low, a relatively neutral pH, absence of obvious biochemical poisons such as cyanide, phenols, or copper salts, and abundance of dissolved O_2. In practice, the minimal diluent flow is five times the flow of effluent, if regeneration within a reasonable distance downstream is to be achieved.

Of all these requirements, the need for O_2 is the greatest. Since the solubility of gases decreases with rising temperature while the rate of biochemical activity often doubles for every 10°C rise in temperature, one may see why the temperature of water entering rivers from industrial plant is so important. Thus a large volume of effluent, even pure distilled water, coming from a power station at, say 40°C, would be disastrous. The effect of increased temperature on river life was shown in England in 1976, when prolonged drought and intense sunlight raised the temperature of many rivers to 25°C, resulting in the death of most fish. Along the S.E. coastline of the U.S.A., the summer heat and thermal pollution from industry regularly raise the temperature of rivers to 33–35°C. All living things die except a few thermophilic bacteria, which are useless for regeneration purposes.

Why is the oxygen so vital? To see this, we must consider how human effluent is dealt with in Western countries. A well-nourished adult excretes about 8 g of combined nitrogen per day. Of this about 6 g is in urea, $NH_2CO.NH_2$, 1 g in uric acid and 1 g, mostly as bacterial protein, in faeces. The latter two can fairly readily be broken down to derivatives that may be re-assimilated by growing bacteria. Urea is quickly hydrolysed by the enzyme *urease* to CO_2 and ammonia, which is toxic to most higher animals, probably because it reduces the activity of the Citric Acid Cycle in the brain. The limit of toxicity for ammonia is low and often surprisingly sharp. Trout, for instance, will die after 24 h in water containing

2 parts per million of NH_4^+ but will survive at a concentration of 1 ppm. Thus the ammonia problem may be solved by dilution (for example when untreated sewage is pumped into the sea), otherwise it must be converted into something less toxic, preferably by biological means. Fortunately there are bacteria which will convert ammonia to the far less toxic nitrite and nitrate. This solution is the one commonly adopted in inland cities of the Western world. The nitrate is very soluble and is removed from the vicinity by streams and rivers.

Conversion of NH_4^+ to NO_3^- is obviously an oxidation. In addition, human beings excrete about 100–150 g of solids each day, much of it as undigested plant polysaccharides (roughage). Microorganisms present in almost any natural water will slowly hydrolyse these polysaccharides to monosaccharides, and if an oxidizing agent is available, will convert the latter to CO_2 and H_2O. This uses a lot of O_2:

$$C_6H_{12}O_6 + 6O_2 \longrightarrow 6CO_2 + 6H_2O \qquad (1)$$

Oxidation of ammonia and carbohydrate are the major chemical processes requiring dissolved O_2 in surface waters. If the oxidations can be carried out, using forced aeration, in an effluent treatment plant, this O_2 demand can be avoided. Unfortunately, treatment plants are seldom completely efficient, and the problem is made much worse if industrial effluents also contain oxidizable organic material. Among the worst offenders are industries that use polysaccharides as raw material, for example sugar refineries, paper or cellulose pulp producers and other users of wood products. These can release very large quantities of soluble carbohydrate into local rivers, to compete with the human effluent for the O_2 brought downstream. What happens if this O_2 supply is inadequate?

In order to answer this question, we need to look at some fundamental biochemistry.

1.2   Terminal electron acceptors

The standard metabolic pathways for glucose breakdown and for the Citric Acid Cycle need not be discussed in detail here; essentially they release free energy locked in the organic molecules which are degraded, and transfer it to *adenosine triphosphate* (ATP), which can be used as a source of free energy by many processes in all cells (e.g. growth and cell division, osmotic regulation). The energy transfers accompany a succession of oxidations ending in the oxidized end-products H_2O and CO_2. All oxidations imply a concomitant reduction; the chemical which is reduced in many oxidations in all cells is the coenzyme *nicotinamide adenine dinucleotide* (NAD). The general oxidation–reduction process may be written:

$$AH_2 + B \longrightarrow A + BH_2 \qquad (2)$$

where AH_2 is to be oxidized (dehydrogenated), and B is the oxidant. In the case of cells we have the more particular equation:

$$AH_2 + NAD \longrightarrow A + NADH_2 \tag{2a}$$

where AH_2 is some oxidizable nutrient or intermediate. The amount of NAD within any cell is very small and finite: it must be continuously re-oxidized, ultimately by an oxidant from the external environment, the reduced product of which can return to that environment. The most familiar oxidant is O_2, with H_2O as the reduced product, but in the present context several other compounds may be of importance.

Oxidation by transfer of hydrogen atoms (as in equations 2 and 2a) may be theoretically treated as a transfer of electrons and protons, thus:

$$AH_2 + B \longrightarrow A + B^{2-} + 2H^+ \tag{3}$$

This enables us to connect this type of oxidation with one involving only electrons, as in

$$Fe^{2+} \longrightarrow Fe^{3+} + \epsilon \tag{4}$$

Oxidants and reductants may then be compared quantitatively with one another in terms of their ability to gain and lose electrons. A table of *standard redox potentials* may be drawn up, based on a comparison with the tendency of hydrogen molecules to lose electrons in standard conditions of acidity (pH):

$$H_2 \longrightarrow 2H^+ + 2\epsilon \tag{5}$$

The redox potential of this system in 1 N acid (that is, when the pH is equal to zero) is taken to be 0.0 volts. However, even in acid waters the hydrogen ion concentration is never as great as 1 Normal, and inside cells $[H^+]$ is usually near that in pure water, namely $10^{-7}M$, which is equivalent to pH 7. Thus biochemists use a secondary standard of redox potential, E'_0, at pH 7. The tendency of hydrogen atoms to lose electrons at this pH can be expressed as a voltage of -0.42 V, while the tendency of O_2 to accept electrons is denoted by a redox potential of $+0.81$ V. Lest any readers think that these 'tendencies' are very abstract affairs, many microorganisms, including the sulphate-reducing bacteria referred to later in this chapter, possess an enzyme called *hydrogenase*, whose function is precisely to catalyse the reversible reaction expressed in equation 5.

The free energy (chemical potential) made available when electrons or hydrogen atoms are exchanged between two redox pairs may be precisely calculated (for details see MORRIS, 1972). This is useful, because it enables the energy released in oxidations to be compared with that needed for energy-requiring (endergonic) reactions, and in particular with the requirement for the synthesis of ATP. This is coupled with redox processes in various complex ways. Such computations provide useful guides to biological possibilities. For example, the (standard) free energy change required for the synthesis of one mole of ATP is 33 kilojoules (kJ). The free energy released by the oxidation of one mole of $NADH_2$ by O_2 is

$$2 \times 96.5 \times (0.82 - (-0.32)) = 220 \text{ kJ}$$

(2×96.5 (Faraday's constant) because *two* electrons are involved); the numbers inside the brackets are the standard redox potentials of the oxidant (O_2) and reductant ($NADH_2$) respectively.

Thus in principle, the oxidation by O_2 of one molecule of any natural substrate whose redox potential is roughly that of $NAD/NADH_2$, could support the synthesis of at least six molecules of ATP. The experimentally determined ratio is three.

Table 1 Biological energy yield of various inorganic oxidants.

	E_0' (mV)	*$\Delta G_0/2\epsilon$ (kJ)
$HCO_3^- + 9H^+ + 8\epsilon \rightarrow CH_4 + 3H_2O$	−230	−8.4
$SO_4^{2-} + 8H^+ + 8\epsilon \rightarrow S^{2-} + 4H_2O$	−200	−9.6
$NO_3^- + 2H^+ + 2\epsilon \rightarrow NO_2^- + H_2O$	+420	−162
$NO_3^- + 6H^+ + 5\epsilon \rightarrow \frac{1}{2}N_2 + 3H_2O$	+750	−344
cf $\frac{1}{2}O_2 + 2H^+ + 2\epsilon \rightarrow H_2O$	+820	−236

* The figures in this column refer to the useful chemical energy (ΔG_0) that can be obtained when 1 mole of the oxidant is reduced under specific conditions.

Substances having a function similar to that of O_2 are called *terminal electron acceptors*. Oxygen itself is used by almost all terrestrial and most marine animals, by plants and by many microorganisms. The release of O_2 during photosynthesis ensures that the vast quantities present in the atmosphere do not sensibly diminish. However, in principle, any electron acceptor which is kinetically active (in practice, one for which an enzymic catalyst has evolved), and which has a standard redox potential more positive than zero volts, can support the synthesis of at least one molecule of ATP per two electrons transferred from $NADH_2$.

Many microorganisms have evolved pathways based on the electron transport chain shown in Fig. 1–1. Dozens of terminal electron acceptors are known, many of them organic aldehydes or ketones, which can be reduced to alcohols. These need not concern us here. Three inorganic acceptors are of importance: $CO_2(HCO_3^-)$, SO_4^{2-} and NO_3^-. They may be reduced as shown in Table 1. The negative sign implies that energy is released when the reaction goes in the direction of the arrow. Because some of the reactions are more complex than others, they have been 'standardized' to show the energy made available per two electrons transferred.

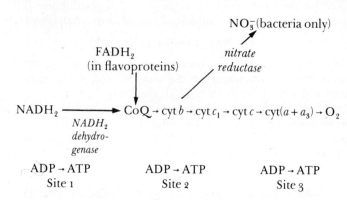

Fig. 1–1 Sequence of electron transport carriers, and of phosphorylation sites. The same basic pathway to O_2 holds true for animal mitochondria and for many microorganisms, but there is more variability in phosphorylation sites in the latter. There are probably two sites for ATP synthesis in electron transport which are coupled to nitrate reduction.

In all the reactions shown in Table 1 hydrogen is the implied reductant, hence the protons and electrons appearing on the left-hand side of each equation. The first two oxidants, bicarbonate and sulphate, appear to be in a class of their own, because the amount of free energy released, per two electrons transferred, is not enough to synthesize a molecule of ATP. Thus it looks as if these two oxidants are biologically not powerful enough. There *are* organic oxidations with redox potentials far below those of hydrogen. An example which will recur in a later section is the oxidation of pyruvate to acetate which may be written

$$CH_3.CO.COO^- + H_2O \longrightarrow CH_3.COO^- + CO_2 + 2H^+ + 2\epsilon \; [E_0' = -0.7V]$$
pyruvate   acetate

The major interest of the first two reactions lies in the products that are formed, rather than in the usefulness of the oxidants. The *methane-forming bacteria* will be referred to in the next Chapter, in connection with biological reactions involving mercury. Moreover, CH_4 helps to recycle carbon from anaerobic aquatic environments where CO_2 is being reduced, to aerobic environments where it is re-oxidized to CO_2, which may be fixed by photosynthetic bacteria or algae (SCHLEGEL, 1974). Attempts (so far not very successful) are being made to treat human sewage in so-called 'dry fermentations', in which methane is an important end-product. One hope is that the yield of the gas will be large enough for it to be an important source of energy to help in replacing fossil fuels. Enormous quantities of methane (thousands of millions of tonnes) are released to the atmosphere every year by biological fermentations (see p. 21).

The *sulphate-reducing bacteria* will be considered in a later section, because of the peculiarly noxious nature of the end-product, hydrogen sulphide.

From Table 1, we see that about as much free energy is gained, per two electrons transferred to nitrate, as would be obtained by the more familiar reduction of O_2. Many microorganisms use NO_3^- as a terminal electron acceptor in the absence of O_2. Frequently O_2 represses the synthesis of the enzyme *nitrate reductase*, so that nitrate is only used in completely anaerobic conditions, but it is then readily reduced. The immediate product is the toxic nitrite, which may diffuse from the cells. Several species of microorganisms will use nitrite itself as an oxidant, some taking the process, by way of NO and N_2O, as far as N_2. (This is *dissimilatory* nitrate metabolism.) Syntropic nitrogen-fixing anaerobes may use the N_2. Other organisms use nitrate in the way plants do; they reduce it, in a controlled fashion, by way of nitrite, to NH_4^+, which is used for intracellular synthesis of amino acids, and hence proteins and nucleic acids (*assimilatory* nitrate metabolism).

Thus an anaerobic body of water may be far from sterile. It can support a large variety of microorganisms using terminal electron acceptors other than O_2. However, the biological activity is limited; no oxygen-requiring animal can live there. More important, it will not support plant life, including phytoplankton, because photosynthesizing organisms generally require O_2 for respiration in order to maintain cell integrity during the hours of darkness. As anaerobic water cannot be re-oxygenated by plants, it must depend on diffusion of O_2 through the air-water interface, which may be very slow. If the organic content of the water is high, even in the absence of anaerobiosis aerobic bacteria will compete for the incoming O_2 with photosynthesizing organisms, and limit the growth rate of the latter.

1.3   Polluted fresh-water streams

A simplified description of a stream which does not receive enough well-oxygenated diluent is that it has a high content of solids, which settle on the bottom as an anaerobic mud, from which bubbles of H_2S and CH_4 may rise. The water will be free from almost all animal and plant life, but a few specialized worms may survive in the sediment. Nitrate arriving in the stream from treated sewage will be immediately reduced; the state of the water will be far worse if untreated excreta are allowed to enter it. Not only is such a stream obnoxious and useless as a source of drinking water, but conditions are favourable for the growth of pathogens such as cholera and typhoid organisms. For this reason officials in many large cities of Europe began, in the mid-nineteenth century, to prohibit the dumping of untreated sewage into rivers, and to develop sophisticated methods of sewage treatment. The 'activated sludge' method was invented in Manchester. The social and financial gain from this development was enormous, and it can still reap benefits today. If such methods of water treatment could be adopted throughout the Middle East they would, for example, eliminate the crippling economic burden of trachoma blindness.

Even the water of a river polluted as heavily as that described in the previous paragraph would become pure many miles downstream, if left to itself. All too often, however, conurbations downstream of the original source of pollution add further burdens before the first has been removed, so that the river remains sterile until it is finally diluted by the ocean (cf. ELKINGTON, 1977).

The aim of pollution control is to prevent the development of such sterility, or to reverse it if it has been allowed to occur. Our expectations must not, however, be too idealistic. They may indeed be fulfilled, as with the Thames and London. Other cities have not had such success.

1.4   The principles of sewage treatment

The fundamental biochemistry of sewage treatment is that oxygen is used to oxidize ammonia, first to nitrite and then to nitrate.

$$NH_3 + 1\tfrac{1}{2}O_2 \longrightarrow NO_2^- + H^+ + H_2O \tag{6}$$
$$NO_2^- + \tfrac{1}{2}O_2 \longrightarrow NO_3^- \tag{7}$$

These are both exergonic processes, carried out by bacteria which derive energy for making ATP from the oxidations. Reaction 6 is catalysed by *Nitrosomonas*; the nitrite is excreted and used as substrate for reaction 7 by *Nitrobacter*. The skill in the process lies in keeping a continuous fermentation of these organisms going as they are both autotrophs, whose growth is inhibited by high concentrations of organic substrates, and also obligate aerobes. Thus there has to be a pre-treatment period, during which solids are removed by sedimentation or centrifugation, while heterotrophic organisms oxidize the organic molecules. During the conversion to nitrate, there is usually forced aeration, to make sure that O_2 is always present in excess.

The ideal sequence is one in which NH_3 disappears completely, to be replaced by nitrite, which in turn is replaced by nitrate as shown in Fig. 1–2. However, there are many things which can go wrong; for example, foaming caused by household detergents can reduce the efficiency of aeration. Thus water treatment engineers are usually happy if the effluent leaves the treatment plant at a point somewhere on the right-hand side of Fig. 1–2, i.e. when most of the ammonia has been converted to nitrate or nitrite.

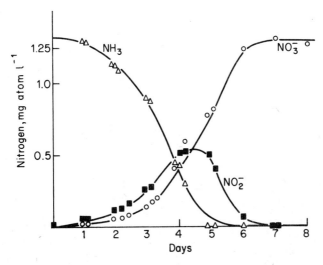

Fig. 1-2 Interrelationship between ammonia, nitrite and nitrate. The symbols show the concentrations of NH_3 ($\triangle$), NO_2^- ($\blacksquare$) and NO_3^- ($\circ$) in water drawn from the Thames estuary in 1960, and incubated for eight days at 30°C in the presence of O_2. The lines are the predicted concentrations of the three chemicals, based on the presence in the original water sample of low concentrations of *Nitrosomas* and *Nitrobacter*, and on the known growth rates of these two organisms. (From KNOWLES *et al.* (1965). *J. Gen. Microbiol.*, **38**, 263.)

1.5 The burden of sewage effluent on rivers

It is desirable to quantify the work that the river has to do before its waters become potable, and this may be done in several ways. A simple test is that of the Biological Oxygen Demand (BOD), which measures the amount of O_2 consumed by a sample of river water in a given time (usually five days). This test does not work very well with heavily polluted water, because there may be a long lag period before conditions are suitable for the growth of *Nitrosomonas* and *Nitrobacter*. A more sophisticated test is to measure the O_2 consumed by the sample over a short period of time under optimal conditions, and to assume that consumption will subsequently fall off exponentially with a characteristic decay constant (see Fig. 1-3). This gives the Ultimate Oxygen Demand (UOD).

These measurements and assumptions apply only to the place from which the sample was actually taken. Clearly, industrial effluent downstream from the sewage works, whither it carries large amounts of fermentable material, or too much waste heat, may undo the best efforts of civic engineers; thus pollution of large rivers has to be considered along their entire length.

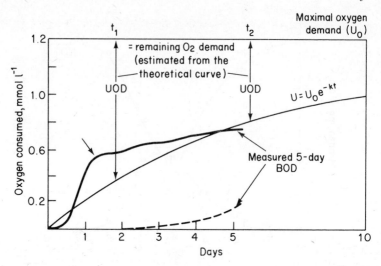

Fig. 1-3 Biological Oxygen Demand (BOD) and Ultimate Oxygen Demand (UOD). The graph shows two possible curves for oxygen uptake against time in a highly polluted waste water. ——— shows the uptake expected if the water contains a readily degradable substrate; at the point indicated by the arrow all the substrate will have been taken up by bacteria, and the slow oxygen utilization thereafter is due to the endogenous respiration of the cells formed in the growth phase. – – – – shows oxygen uptake in water containing a slowly degradable organic substrate, or effluent from a treatment plant in which the NH_3 has only partially been converted to NO_3^-. Note that the 5-day BOD test might grossly underestimate the true oxygen requirement. ——— shows how UOD may be estimated (for example, at t_1 and t_2) from a theoretical curve. This method can only be used if the oxygen demand at infinite time (U_0) can be estimated by some means.

1.6 Nitrate and nitrite

The best modern sewage works may reduce the nitrogen output of their effluent by as much as 40% by using denitrifying bacteria to convert NO_3^- to N_2 (Table 1). It is, nevertheless, inevitable that the mineral content downstream of a centre of population is higher than that upstream. Sodium, potassium, phosphate and inorganic nitrogen compounds are all nutrients which may cause troublesome overgrowth (eutrophication) in rivers, lakes or coastal seas. However, nitrate and nitrite pose a unique problem in addition to their role as nutrients.

Nitrite is toxic to man and many animals, and is likely to be produced in an anaerobic environment by many organisms, according to the equations of Table 1. The most widely known poisonous effect of nitrite is its ability to cause methaemoglobinaemia. Nitrite oxidizes the ferrous iron in the haemoglobin of the body's red blood cells to ferric iron:

$$(Hb)Fe^{2+} + NO_2^- \xrightarrow{\quad} (Hb)Fe^{3+} + NO_2^{2-} \xrightarrow{\;2H^+\;} H_2O + NO \qquad (8)$$

$(Hb)Fe^{3+}$ represents *methaemoglobin*, which is unable to transport oxygen in the blood. The victim suffers from anaemia which can be fatal. Infants are most at risk from this, partly because their haemoglobin is more easily oxidized, and partly because nitrate reduction can occur readily in the anaerobic, only faintly acid conditions of a baby's stomach. The secretion of hydrogen chloride by the stomach develops very slowly during the first six months of life.

Adults are much less at risk from nitrite-induced methaemoglobinaemia, but there is also the danger of nitrite reacting with certain aromatic amines, such as food additives, in the gut to form compounds called *nitrosamines*. These contain the grouping

$$\begin{array}{c} R-NH \\ | \\ N \to O \end{array}$$

Nitrosamines are carcinogenic, and there is good evidence to connect them with the development of cancer of the stomach lining.

Thus even the inflow of completely treated sewage into river water places a biological time-bomb in the environment. If the river becomes anaerobic anywhere in its subsequent course, nitrate will be partially reduced to nitrite in the stream. In addition, nitrate itself is potentially dangerous to babies because their intestinal flora can reduce it to nitrite. High nitrate concentrations are more likely to occur in the summer months, when diluent inflow is low. In the great English drought of 1976, the nitrate content of the tap water in several cities rose above the safe limits for babies, and supplies of pure water from springs were stockpiled for issue. It is fairly certain that within the next ten years, the issue of nitrate-free water for making-up infant feeds will be routine in the London area. It will not be easy to find water on this scale without distillation, because nitrate, like other inorganic ions, can percolate down to the underground aquifers from which pure water has hitherto been drawn. This nitrate often comes from excessive use of chemical fertilizer.

1.7　Other aspects of sewage disposal

Attention has been concentrated on the problem caused by human nitrogenous excreta, although pollution of surface water by ammonia, nitrate and phosphate, derived from leaching of excess fertilizer applied to agricultural land, can also be a serious problem. In some areas it is estimated that up to half the nitrogen content of a stream may come from fertilizer run-off.

It is nevertheless right to think of people as the more fundamental problem, because over-use of fertilizers can be cured by education or rising costs. It is interesting that the population centres with the worst pollution problem, such as Calcutta, are those where farmers can least afford to be wasteful with fertilizer. The only solution to the sewage problem is for there to be fewer people. In effect, the use of rivers as a depository for human and industrial effluent, even with enlightened biological treatment, creates a problem that cannot be cured, quite apart from the continuous desperate search for increasing amounts of oxygenated diluent water. Civil engineers have hardly begun to think about the problem (see HANLON, 1977). For example, biological recycling, (e.g. by using fishponds) is not a solution on the urban scale. Fishponds are shallow ponds – so that the water at the bottom does not become de-oxygenated – whose high rate of nitrogen input from sewage supports a vigorous growth of aquatic plants and hence of fish which are netted at intervals. BONEY (1975) gives a more detailed discussion. The technique makes sound biochemical sense, but leads to dangers of endemic disease, especially by parasites.

There are other serious chronic freshwater pollutants besides nitrogen. Chief among them is phosphate, which comes both from breakdown of metaphosphate water softeners of the 'Calgon' type and from leaching of agricultural fertilizer.

1.8   Sulphate and sulphide

Sulphate-reducing bacteria are exacting anaerobes, with several odd, perhaps primitive, properties. Their wide distribution, and tolerance of extreme environments, makes it important to discuss them briefly.

The overall energy yield of SO_4^{2-} reduction has been shown in Table 1, but in fact the activation of the rather stable sulphate ion and its reduction to SO_3^{2-} is energy-requiring, so that these organisms might be thought of as sulphite-reducing. This reminds us first that they will use many sulpho-ions other than sulphate as terminal electron acceptors, and second, that sulphite, used in the purification of wood pulp for paper-making, is an important pollutant of fresh-water environments. The overall reduction may be written

$$ATP + SO_4^{2-} \longrightarrow APS + pyrophosphate$$
$$\text{(adenosine phospho-sulphate)}$$
$$APS + 2H^+ + 2\epsilon \longrightarrow AMP + SO_3^{2-}$$
$$SO_3^{2-} + 6H^+ + 6\epsilon \longrightarrow S^{2-} + 3H_2O$$
$$S^{2-} + 2H^+ \longrightarrow H_2S$$

The electrons and protons needed for these reductions usually come from one of two sources; one is molecular hydrogen (see p. 4). As Table 1 shows, this does not provide much useful energy. The more common source of reducing power is the oxidation of lactate:

$$CH_3.CHOH.COO^- \longrightarrow CH_3.CO.COO^- + 2H^+ + 2\epsilon$$

lactate   pyruvate

$$CH_3.CO.COO^- + HPO_4^{2-} \longrightarrow CH_3.CO.OPO_3^{2-} + CO_2 + 2H^+ + 2\epsilon$$

acetyl phosphate

$$CH_3.CO.OPO_3^{2-} + ADP \longrightarrow CH_3.COO^- + ATP$$

acetate

The standard redox potential for pyruvate oxidation is very negative (p. 6).

The acetate which is formed is excreted, while both lactate and hydrogen are provided by other microorganisms. Thus the sulphate-reducing bacteria are, necessarily, symbionts. If the medium is illuminated, still other organisms may use H_2S as hydrogen donor in anaerobic photosynthesis. Such a complex of co-existing organisms has been called a 'sulphuretum'.

One potentially valuable property of H_2S production is that it forms insoluble sulphides with almost all heavy metal ions. The mud of industrial estuaries may contain high concentrations of metals such as zinc or copper, which have been precipitated from solution by H_2S, while the lower depths of the Black Sea are rich in suspended iron sulphide. Apart from this, hydrogen sulphide is very undesirable indeed. It inhibits cytochrome oxidase, the terminal oxidase of aerobic organisms, even more strongly than hydrogen cyanide, so that very few aerobic animals and plants can live in water containing H_2S. Even in less than toxic concentrations, its smell and the taste of its solution are highly objectionable. Finally, it is very corrosive, especially towards iron, and particularly if the exposure is intermittent.

1.9   Lakes

The use of lakes for drinking water, and the disposal of human and industrial effluent, is no different in principle from the use of river water, and it might at first sight seem that the greater volume of water available for dilution would make pollution less formidable. There is, however, one crucial difference between rivers and lakes. The pollutants of a river will eventually flow to some unfortunate community downstream: as the Swiss export their pollution to the Germans, and the latter, with the French, burden the Dutch, by means of the Rhine as it flows from the Alps to the North Sea. The results of mismanagement of lake water stay where they are, with the community that made them. Thus Lake Erie is dead; it is unsafe to swim in, or even to sail over, but it remains obstinately outside the doors of the citizens of Detroit, and the other lakeside towns that made it so.

A lake is, in effect a trap which retains almost all the solid and dissolved matter that flows into it. Geologically speaking, lakes are temporary phenomena and they are sooner or later filled up, as the enormous sedimentary deposits of the Holocene bear witness.

There are, however, great differences in the rate at which infill takes place. Hydrologists distinguish between *oligotrophic* lakes, which lie in hardrock districts, get little soluble matter from the rocks or vegetation surrounding them, and consequently can support only a low density of plant and fish life; and *eutrophic* lakes, lying in softer rock surrounded by a more fertile soil (and usually a denser population).

The increase in the productivity (fertility) of lake water is called *eutrophication*. Human populations can speed this up to such an extent as to diminish drastically the usefulness of the lake water, and the process is almost irreversible. It is compounded by the peculiar behaviour of large volumes of water.

Lakes, unless very shallow, rarely freeze completely even when the surface temperature goes below zero for long periods, because the maximum density of water lies at 4°C, so that colder water floats to the top. The body of water which lies below the frost boundary in the surrounding soil remains at 4°C. When the air becomes warmer in the spring this lower part of the lake does not, as might be expected, become warmer. Instead a sharp temperature boundary, the thermocline, is established about 16 m down; above this level the water circulates by convection currents and is rapidly warmed by sun and air. The water in the lower layer, or *hypolimnion* (Fig. 1–4) remains at 4°C and does not physically mix with the upper layer, or *epilimnion*. At 16 m there is very little light, even in the clearest lake, for photosynthesis, and in consequence the hypolimnion only becomes oxygenated at the two seasons in the year (spring and autumn) when the temperature of the epilimnion is also 4°C, and bulk mixing can take place. Winds, by turbulence, increase the depth at which the thermocline is found, and blur the boundary, producing an intermediate zone, or *metalimnion* (see BONEY, 1975). The persistence of strong winds over the open oceans helps to prevent the establishment of permanent temperature boundaries, except in the tropics.

The charge of oxygen that the hypolimnion receives in spring has thus to sustain all the aerobic activity within it, both plant and animal, until the autumn. At the same time, the hypolimnion must deal with the rain of solid matter brought in by the inflowing rivers, and also with the organic matter released on the death of organisms in the epilimnion. This will happen faster in summer, when higher temperature and abundance of sunlight provide favourable conditions. If the supply of nutrients to the epilimnion becomes too great, it is the hypolimnion which first becomes anaerobic, with serious consequences.

The nutrients required for growth and replication of organisms are nitrogen, which, as we have seen, is continuously added to rivers by human activity, together with potassium, magnesium and phosphate, and many trace elements. It is often phosphate which is the limiting ingredient (see Table 2, p.18) and the amounts added to lakes over the last

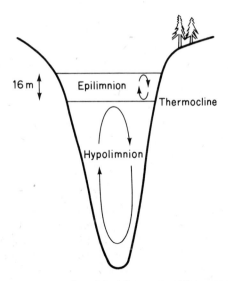

Fig. 1–4   Temperature boundaries in lake waters. The separation of the two water masses, each with its own convection circulation, occurs because of the density maximum of water at 4°C. In winter the epilimnion will be colder, and in summer warmer, than the hypolimnion. The temperature boundary, or thermocline, is about 16 m below the surface in still water. Winds blowing above the surface will push the boundary deeper, and also make it less sharp.

few decades from washing powders and fertilizers can make a serious difference to the biological balance. In hot summer weather when the dissolved oxygen concentration is low, the epilimnion of lakes can become anaerobic; in which case shallow-water fish die. Algal blooms may form, and reduce the attractiveness of the lake waters for leisure purposes. Moreover, by-products of some algae are toxic to cattle. The water may become unpalatable for human consumption, and algal remains clog the input filters of the water purification plants.

However, anaerobiosis in the hypolimnion is more serious even than this. Debris which passes through the thermocline sinks to the bottom as a mud similar to that in a polluted river, and the nutrients in the mud support the activities of a wide range of anaerobic bacteria. Less dense products (H_2S, CO_2, CH_4, and sometimes other specialized pollutants) rise, perhaps as bubbles, through the water column back to the epilimnion. The inorganic nutrients released on the death of the bacteria remain in the mud and are taken up by new microorganisms. When a lake has reached this stage, in which both epi- and hypolimnion are anaerobic for much of the year, it is said to be 'eutrophicated'. The lake waters, previously used for drinking water by the towns on the bank, become foul-tasting, and if there is an effluent stream, it is useless as a supply of diluent for the reasons outlined in the previous section.

It is, unfortunately, almost impossible to regenerate a eutrophicated lake. Even if all pollution is prevented from entering its waters, the mud on the bottom has to be removed, which has only proved possible for small lakes. In the neighbourhood of Geneva, which depends heavily on lake water, as much as a million tonnes of oxygen per year has been pumped into the hypolimnion. This treatment has been successful, but is obviously a desperate and expensive remedy. In general, it is only possible to cut down the amount of nutrient entering the lake (particularly phosphate), and to hope that in a few decades aerobic conditions will be re-established. Some hope is provided by the partial regeneration of Lake Michigan, but Lake Erie, which is only 50 m deep, is still sterile. This is also the present state of many Swedish and Finnish lakes, eutrophicated by untreated sewage and wood pulp effluents. Further east, strict control is yet to come. Immense amounts of nutrients brought in by the Volga are rapidly eutrophicating the Caspian Sea; in addition the bottom of the southern half of this sea is said to be covered by several centimetres of tar from the oil fields of Georgia. The floor of Lake Baikal, the largest fresh-water lake in the world, is reputed to be covered by millions of tonnes of waterlogged tree trunks, lost during lumbering operations, which will provide enough oxidizable carbohydrates to keep the bottom mud anaerobic for centuries.

2 Marine Pollution

2.1 Replacement times

Eutrophication applies with just as much force to enclosed seas as it does to fresh-water lakes. How complete it is depends on the salinity of the inflowing water, as well as on the replacement time for the body of water itself. For the North Sea the replacement time is two years, and there is little danger that the lower layers will ever become anoxic. The same is true of the deep Mediterranean (not the coastal areas or the Adriatic); although replacement time is 100 years, it shows very rapid mixing, at least in spring. By contrast, although the replacement time for the Baltic is only 35 years, mixing is poor because very little salt water comes in, and 60 000 km² of the southern Baltic is now a 'marine desert'. The Gulf of California and some Norwegian fiords show similar effects. Finally, the Black Sea, whose replacement time is 2000 years, is completely and permanently anaerobic below 180 m, and the hypolimnion contains H_2S, the concentration increasing with depth (max. 2000 m). Unlike the Baltic, which has become eutrophicated within the last generation (it is thought largely because of agricultural phosphate run-off brought in by northward-flowing rivers), eutrophication of the Black Sea can have owed little to man's intervention.

The tendency is to assume that, by contrast, the open oceans provide an inexhaustible reservoir for pollutants. To a degree, this is true. For example, the accumulation of nitrate and phosphate in the deep oceans is far greater than that in shallow seas and lakes (Table 2), and this water is continuously brought to the surface in many upwelling regions; by comparison, enrichment of *ocean* water by run-off from the land is trivial. However, even the great oceans may be affected by sub-lethal concentrations of pollutants. Low concentrations of dimethyl mercury (p. 21) and organochlorines (Chapter 4) are known to have deleterious effects on phytoplankton. This might conceivably reduce the yield from ocean fisheries. The global marine phytoplankton growth rate exceeds the fish harvest by a factor of perhaps 1000-fold, but fish can only be caught in large numbers in a relatively small area of the oceans. A small drop in the efficiency of plankton growth in these areas might drastically reduce the harvest. Another cause for concern is the possible concentration in the food chain, of substances originally present in very low concentrations. Some algae have the capacity to concentrate elements, such as iodine, to an astonishing degree. So far no evidence of significant and persistent accumulation of radioactive waste products has come to light, but the possibility is an uncomfortable one.

Table 2   Nitrate and phosphate in marine and fresh-waters (values in microgram-atom per litre).

	Nitrate-N		Phosphate-P	
	max.	*min.*	*max.*	*min.*
Eutrophic fresh-water (Lake Windermere, England)	28.5	7.0	0.17	0
Shallow sea (English Channel)	10.5	0	0.7	0.03
Deep ocean (Pacific, >1000 m)	36		2.9	

Since marine waters are not required for drinking, or as a reservoir of dissolved oxygen, in the same way as fresh waters, even marked pollution of enclosed seas is not regarded so seriously, although the disappearance of deep-water fish is unwelcome. However, around the shores of these seas pollution may be very unpleasant indeed, as Fig. 2–1 indicates. The chief effect is on holiday-making, which is a very large industry, but littoral food species may also be badly affected, although it must be remembered that this may be due to contamination with pathogenic bacteria rather than chemicals.

Only two pollutants will be considered in detail in this chapter – mercury and oil. The former affects directly only shallow bays and estuaries, but may also affect fresh-water lakes. The latter can occur anywhere in the oceans; it introduces a biochemical topic of great importance (mixed-function oxidation) that will be referred to frequently in later chapters.

2.2   Mercury

Elemental mercury (Hg) is not very poisonous by mouth; an adult might tolerate 30 g day^{-1}. The vapour is more toxic, giving rise, over a long period, to hatter's shakes. The toxicity of inorganic mercury salts is related to their solubility; the poorly-soluble calomel (Hg_2Cl_2) was long used as a purgative. Mercury ions form strong complexes with the -SH groups of proteins; their toxicity is probably due to inactivation of proteins in cell membranes. This seems likely as effects are particularly noticeable in kidney and brain, both organs in which membrane function is important, and because many bacteria and fungi are killed by mercury compounds. Unspecific bactericidal activity is often attributable to damage to the cell membrane.

Inorganic mercury salts have now been totally replaced as drugs, fungicides, bactericides, etc., by so-called *organic mercurials*. We are accustomed to thinking of metals as elements which form only ionic compounds (salts), but many metals also form compounds with covalent bonds (shared electrons). Tin (p. 38) and lead (p. 32) compounds will be mentioned later in this book. Mercury readily forms covalent bonds,

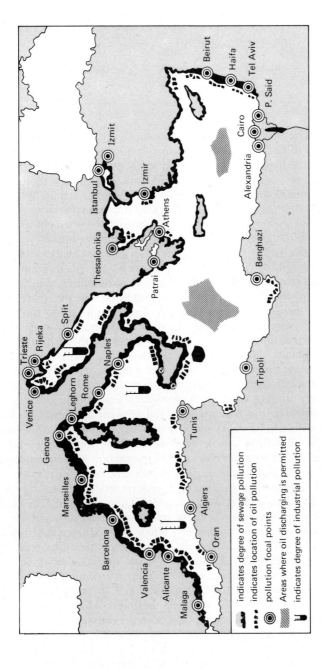

Fig. 2–1 The distribution of pollution around the Mediterranean. (Redrawn from *The Sunday Times*, 1 June, 1975.)

especially with aromatic compounds, and several of the resulting compounds, such as the one shown below, are used as fungicides and

pesticides. The advantages of such compounds are that the solubility can be controlled by the inclusion of suitable substituents on the benzene ring, while the benzene-Hg bond is so stable as to provide, in effect, an R.Hg$^+$ ion, which is still able to react with -SH groups to form an [R.Hg.S–protein] derivative. Insoluble compounds, such as 'Semesan', have been widely used for impregnating seed corn to protect it against soil pests. Fatalities, sometimes numerous, have occurred when peasants, not warned of the dangers, have used treated corn to make bread.

This was the background knowledge when a disease broke out in 1953 in a Japanese fishing village, Minamata, which has given its name to the disease. The symptoms were observable both in humans and in animals, and were primarily neurological – numbness in the extremities, locomotor disorders, convulsions. A few people died, but more frighteningly, there was a marked increase in the number of deformed children born in the village. The causative agent was therefore a *teratogen* as well as a neurological poison. A teratogen is a substance which interferes with the orderly development of the foetus; it is not necessarily a mutagen or carcinogen. Since the disease neither died away nor spread into the surrounding countryside, medical scientists concluded that its origin must be in the environment, and that the agent was probably absorbed with the food. As most victims were fishermen, this turned attention to the shallow bay on the shores of which the village stands.

When analysing the anaerobic ooze on the sea bottom, scientists found a phenomenally high concentration of mercury, much of it as the free element. Values as high as 2.1 g kg^{-1} were recorded. The source of the pollution was a plastics factory, using Hg$_2$SO$_4$ as a catalyst. The factory had been operating very inefficiently; it was estimated that it had released 600 tonnes of mercury into the bay. When this discharge was stopped, 'Minamata disease' slowly abated. There were other outbreaks, for similar reasons, at various locations in Japan, but none so serious. The scientific puzzle, however, remained – no mercury compound was known to cause this combination of symptoms, and in particular, metallic mercury was thought to be relatively harmless. How could mercury pollution, no matter how severe, have been the causative agent?

The solution to the problem was found in Sweden some 15 years later (JENSEN and JERNELØV, 1969). There had been a marked decrease in the bird population around the lakes of central Sweden; fish-eating birds and raptors were particularly affected. The tissues of dead birds contained surprisingly high levels of mercury, but the nature of the diet ruled out poisoning from mercurial seed dressings. Industrial pollution from the factories around the lake shores, making products from wood pulp, was then suspected. Mercurial fungicides, added to the pulp to preserve it, at first seemed likely culprits after concentration in food chains, but later mercury itself, released accidentally from plants making caustic soda, was thought to be the cause. So far the story was parallel to that in Minamata, but this time a new tool, the gas chromatograph, was available. With its aid scientists were able to show that the tissues of the birds – and of fish in the lakes – contained significant traces of a mercury derivative hitherto unknown in biological environments. This was *dimethylmercury* ($CH_3.Hg.CH_3$).

This covalent compound had been synthesized many years previously, and was known to be a powerful neurological poison, but there had been no reason to suspect its existence in the environment. Further studies established that it was indeed present in the lake water, and also the mode of its formation. Like many nerve poisons (see Chapter 4), dimethyl-mercury is very lipid-soluble. It fairly readily loses a methyl group to form the ion $CH_3.Hg^+$, which is also toxic. Both compounds are slightly soluble in water.

Anaerobic bacteria in the bottom mud were found to be responsible for the conversion of Hg metal, or inorganic salts of the element, to dimethylmercury. As mentioned in the previous chapter, there is a group of bacteria which reduce CO_2 to methane (Table 1), and which are responsible for the enormous quantities of methane liberated into the atmosphere each year. The final stages of the reduction process in such organisms involve a coenzyme which contains an atom of cobalt, and is sometimes called *cobalamin*. It is a close relative of the vitamin, B_{12}. The relevant stages of the process are shown in Fig. 2–2. It has subsequently been found that anaerobic bacteria can readily oxidize mercury metal to Hg^+, and can just as readily release inorganic mercury from organic complexes such as that shown on p. 20.

The consequence of the powerful and unspecific activities of the benthic anaerobes is that, once the environment has become polluted, a mercury cycle is set up which is very difficult to break. It is therefore very important both to monitor mercury losses from industrial sources, and to improve the general quality of the aqueous environment, whether marine or fresh. The less anaerobic mud there is on the bottom, the fewer methanogenic bacteria will be present. In heavily polluted areas, clean-up is, unfortunately, a long-term process. In the open sea, the situation might be expected to be different. There is an almost unchanging concentration of Hg ions in seawater, which presumably come from breakdown of rocks; Fig. 3–6 (p. 37) shows that the pattern of Hg deposition in sediments is different from that of such recent pollutants as lead and cadmium. Since the ocean bottom is not anaerobic, there ought

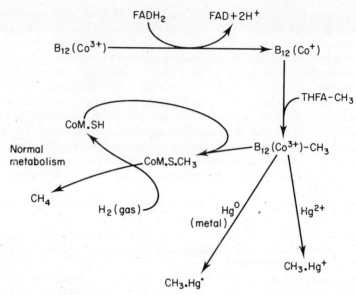

Fig. 2–2 Schematic diagram of mercury methylation. THFA–CH₃ stands for methyl-tetrahydrofolic acid. Tetrahydrofolic acid is a coenzyme that carries a carbon atom in the cell during its reduction from CO_2 to CH_4. The cobalamin coenzyme (p. 21) is written B_{12} for the sake of brevity. CoM is the newly-discovered 'methylation coenzyme' of bacteria – $CH_3.S.CH_2.CH_2.SO_3^-$.

to be no significant dimethylmercury formation in the deep sea. However it has occasionally been found that the oily muscle of tuna, a fish of the open oceans, contains Hg in concentrations above 1 μg g^{-1}, the recommended limit. This is almost certainly because tuna is carnivorous, and accumulates dimethylmercury from its prey.

2.3 Oil pollution

The *Amoco Cadiz* disaster, in which a tanker was wrecked off Brittany in 1978, releasing over 200 000 tonnes of crude oil on to the coast, is only the most dramatic in a series of large-scale accidents which are almost inevitable, given the present scale of sea-borne traffic. However, spillage into rivers, or the daily incidents in which ships legally or illegally release small volumes of oil into the sea, produce an annual pollution rate several times greater than that due to shipwrecks. Regions undisturbed by currents, such as the Sargasso Sea, are said to be covered by a mass of floating 'cakes' of crude and fuel oil emulsified with seawater. In time this oil disappears; it is only near an inhabited coastline that the question of the *rate* of removal becomes important. Given the chemical inertness of many petroleum hydrocarbons, this is a considerable achievement for the marine ecosystem. How is it done?

In purely physical terms, about one third of a typical spillage of crude oil will be volatile enough to evaporate fairly quickly. The rest will be a viscous liquid, possibly containing toxic components such as phenols, which will asphyxiate or smother any surface life, such as seabirds or algae, which it covers. Left to itself the oil will incorporate droplets of water to form a stiff emulsion (called a 'mousse'), the major component of the 'tar' found on beaches. The oil, or mousse, can be sunk to the sea bottom by sprinkling some inert solid on it, but, biologically, this is not a very wise thing to do. Alternatively, it is possible, by artificial means, to make an oil-in-water emulsion (like milk) in which the oil is broken up into tiny droplets, by using a surface-active agent to lower the surface tension of the oil–water interface. The agent should preferably be one which causes the droplets to repel one another, thus keeping the emulsion stable.

Detergents have been used in large quantities for dispersing oil slicks, with some success, but they are mostly water-soluble, with strong ionic charges for mutual repulsion of the droplets, and are biologically rather stable. As detergents concentrate at any lipid–water interface, such as a cell membrane, they do a lot of damage to the environment. (Detergents with positive ionic charges, such as *cetrimide*, are widely used as bactericides.) Taking the long-term view, therefore, unless the oil is adsorbed to the rocks and sands of the intertidal zone, where it may stay indefinitely unless treated, the ecosystem may be better preserved if the use of detergents is avoided; although newer compounds, perhaps soluble in the oil phase and without ionic groups, may be less damaging.

In seawater there are many bacteria which will oxidize paraffin hydrocarbons (alkanes), and recent research suggests that the genetic coding for the enzymes that enable them to do this may be transmitted from one bacterium to another by means of plasmids. It should be emphasized that the concentration of such bacteria in unpolluted seawater is low, and that their effectiveness may be limited by the need for other nutrients. Tri-octyl phosphate as a source of phosphorus, and urea dispersed in paraffin as a source of nitrogen, have been successfully sprayed on oil to increase growth rates; both nutrients are oil-soluble, and thus not lost in the sea.

2.4 Mixed-function oxidases

A set of enzymes which will oxidize paraffins will now be described. This is important not only for the bacterial oxidation of alkanes, but also for many oxidations in animal tissues that will be referred to in following chapters. Oxidation often means the removal of hydrogen atoms, or even simply of electrons, from the substrate, as in Chapter 1. Here, however, it really does mean the addition of an oxygen atom, which comes from atmospheric O_2. A great deal of research has been done on this reaction, and it is now clear that there are two classes of direct oxidations: (1) those in which *both* atoms of a molecule of O_2 are added to the substrate; and (2) those in which only *one* atom is added. The first class (*dioxygenases*) are important in the breaking open of the aromatic rings of such pollutants as phenol, but will not be further referred to here.

With enzymes of the second class we may immediately ask, what happens to the other atom of oxygen from a molecule of O_2? Isotope studies have proven that it ends up in H_2O, and this implies that two hydrogen atoms have come from somewhere to reduce it. Metabolic studies show that a reducing agent such as $NADH_2$ or $NADPH_2$ has to be present, and is oxidized concurrently with the substrate. The term *mixed-function oxidase* was coined to describe the overall process. We may also ask, how is the O_2, normally a rather stable molecule, activated sufficiently for it to be split into two atoms? This question was solved by the discovery of a special haemoprotein, called cytochrome P_{450}, which is widely distributed in animal cells and in bacteria. The complete process is shown in Fig. 2–3.

Alkanes are usually attacked at the terminal -CH_3 group (this is called ω-oxidation), to give the product $R.CH_2OH$, which can then be oxidized to the carboxylic acid, $R.COOH$, by known enzymes. Carboxylic acids are substrates for the pathway of β-oxidation. There are other possible

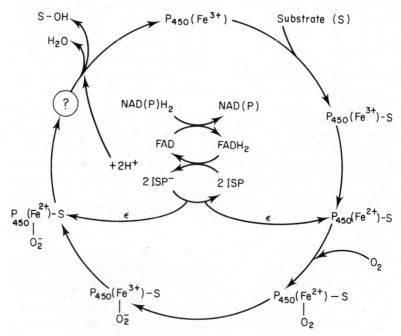

Fig. 2–3 The mechanism of mixed-function oxidation. S stands for the substrate, ISP stands for iron-sulphur protein, or ferrodoxin, one of a group of proteins in which iron is complexed with -SH groups rather than haem. The iron is able to change valency readily.

oxidation mechanisms in bacteria, perhaps involving peroxy compounds ($R.CH_2OOH$), almost all begin with ω-oxidation. Bacteria oxidize alkanes surprisingly rapidly, but they have first to take the substrate, as a lipid droplet, into their cytoplasm, and this may be a rate-limiting process with lumps of tar. Clearly both the initial mixed-function oxidation and the subsequent β-oxidation are oxygen-dependent processes, the chief reason why sinking oil spillages to the sea-bed, where O_2 supply is limited, is unwise. Another reason is that bivalves, which generally live on the sea floor, possess only a trace of mixed-function oxidase activity, at least towards aryl hydrocarbons, unlike crustaceans and vertebrates.

We shall see in later chapters that the mixed-function oxidase systems are remarkably unspecific with respect to their substrates. A wide range of alkyl and aryl (aromatic) hydrocarbons may be hydroxylated. Some of these are naturally occurring compounds, such as steroids, but others are foreign chemicals – drugs or pollutants. In addition, mixed-function oxidases may add an oxygen atom to a N or S atom in an organic compound, or even replace $=S$ by $=O$ (desulphuration).

Not all these substrates are attacked by a single enzyme, however unspecific. Several catalyst proteins (perhaps 5–6) have been identified and two slightly different cytochromes. More importantly, for present purposes, is the existence of two classes of inducible mixed-function oxidases (at least in higher mammals). The MFO systems belong to the small class of mammalian proteins whose concentration fluctuates rapidly in response to inducers. It has been found that the class of oxidases whose typical substrate is the drug barbiturate is induced by this and similar substances, which, however, have no effect on the other class of oxidases. This class, known as the PAH oxidases, attack polycyclic aromatic hydrocarbons, and a typical inducer is benzpyrene (p. 36). The position is complicated by the fact that many inducers, such as the organochlorine compounds discussed in Chapter 4, are not substrates.

Mixed-function oxidases are usually cytoplasmic membrane-bound enzymes. They are found in high concentrations in vertebrate liver, but also in other tissue such as lung. In invertebrates they will be found in hepatopancreas or fat body.

3 Air Pollution

Almost all chemicals released into air eventually find their way into soil, often being washed down by rain. Chemicals such as pesticides (dealt with in Chapter 4) are often sprayed as mists which may drift for several kilometres from their intended point of application. There is, however, a group of pollutants which are gases, or are first released into air, either deliberately to disperse them, or accidentally. Many of these chemicals have a particularly harmful effect on the lungs. Unfortunately, the information available on these pollutants is limited, for two reasons:

(1) There is little *biochemical* information about many of the most dangerous aerial pollutants, particularly asbestos fibres, which can cause virulent lung cancer (mesothelioma), sometimes many years after exposure. The harmful concentration of fibres in air is very low indeed. Asbestos is still widely used, and it is not clear to what extent the general urban population is at risk. The same lack of knowledge applies to the nitrogen oxides from car exhausts which form the irritant component of 'automobile smog', and, in a more limited environment, to nickel- and chromium-containing dusts which can initiate lung cancers.

(2) Turbulence in the atmosphere distributes fallout from volcanic and nuclear explosions all over the Earth's surface. Even circumpolar snows contain traces of radio-strontium and general pollutants such as lead and chlorinated hydrocarbons. Nevertheless, most atmospheric pollution is a rather local affair; there is a vast number of chemical processes each of which could conceivably pollute the atmosphere around a factory using that process. It is therefore difficult to know where to draw a line, but as far as possible, only topics that are of general concern have been chosen for discussion.

3.1 Carbon monoxide

At first sight it is odd that the possible accumulation in the atmosphere of carbon monoxide (CO), which is extremely toxic, does not cause concern while CO_2, which is only harmful in very special circumstances, should alarm doom-watchers (because of the possible 'greenhouse effect' – absorption of long-wave solar radiation by CO_2 may warm up the atmosphere).

There is in fact a carbon monoxide equilibrium (SEILER, 1974); at the earth's surface the atmosphere receives about 6×10^8 tonnes year^{-1} from man-made sources (chiefly automobile exhausts), and perhaps 1×10^8 tonnes year^{-1} from bacterial and algal metabolism in the upper layers of the sea. On the other hand, many bacteria (but probably not plants) oxidize CO, methane-oxidizing organisms by means of a mixed-function oxidase, while many methanogenic and sulphate-reducing bacteria can carry out a process summarized by

$$4CO + 4H_2O \longrightarrow 4CO_2 + 8H^+ + 8\epsilon$$

These exchanges are much smaller than the production in the troposphere from oxidation of methane ($1-4 \times 10^9$ tonnes year^{-1}, balanced by oxidation of the CO to CO_2. Both processes use free OH radicals, which come from the photodissociation of water vapour. Thus the global turnover time of carbon monoxide is less than one year.

Carbon monoxide is, nevertheless, a very serious local pollutant, formed whenever carbon is burnt with an insufficient supply of O_2. A concentration of 10 ppm will produce symptoms of poisoning, especially over a period of time; concentrations many times greater than this are often found in city streets, especially those fronted by high buildings. Most modern automobiles do not produce appreciable quantities of CO, but a few – and most heavy goods vehicles – are very bad offenders, especially in low gear.

Carbon monoxide inhibits many haem-containing enzymes (e.g. cytochrome P_{450}, p.24), but it is the interference with oxygen transport in the blood that causes illness or death. All tissues are affected, but it is the failure of brain function that is decisive. Carbon monoxide binds to haemoglobin 250 times more strongly than does O_2, and so displaces the latter. The binding of both gases is reversible, so that the equilibrium in favour of Hb–CO can be reversed if the O_2 concentration in the lungs can be made high enough, and if the sufferer is still breathing.

3.2   Sulphur dioxide

Both SO_2 and SO_3 are strong irritants in their own right, the latter probably because it combines readily with H_2O to form a mist of H_2SO_4 droplets. Sulphur dioxide is more reactive, and is widely used as a bactericide and fungicide, that is, it has cytotoxic properties quite apart from the acidity of its solutions. The reasons are obscure; SO_2 forms addition compounds with a number of reactive metabolites, but this does not necessarily cause cell death.

The oxidation of SO_2 to SO_3 is catalysed by nitrogen oxides, thus the unpleasant effects of automobile smog are due to a mixture of SO_2, SO_3, NO_2, and ozone (the latter formed when NO_2 and O_2 are mixed in sunlight). This is why new regulations for car exhaust gases are so stringent with respect to nitrogen oxides. All four gases attack sensitive cell layers exposed to air – the conjunctiva of the eye, and, more especially, the epithelium lining the alveoli of the lung. Specialized cells in the latter produce mucous and serous secretions to try and wash away the irritant, and these reduce the efficiency with which oxygen diffuses through to the haemoglobin in the red blood cells in the pulmonary circulation. The effects are most serious in bronchitics, who are already suffering from impairment of respiratory function.

Severe development of smog conditions demands an intense source of the sulphur oxides, and a stable layer of still, moist air at ground level which hinders their dispersion. The first condition requires a highly-populated (urban) environment. The second is often aided by the formation of a 'temperature inversion'.

The automobile is not the major source of atmospheric pollutants. The private demand for energy and heat is the main culprit, especially in temperate regions. Until the late 1950s, in Britain this demand was mainly supplied by coal; the annual emission of pollutants was 8 000 000 tonnes a year of soot, tar and sulphur oxides, with similar rates of emission elsewhere in the world. For example, in the neighbourhood of industrial plants on the northern shores of the Caspian, the soil can contain 6 g kilo^{-1} of tars. Analysis of sediments (see Fig. 3–6) has confirmed that the major source of pollutants has been coal, rather than oil, or factory emissions. About 40% of annual SO_2 released into the atmosphere in the northern hemisphere comes from smoke; the rest is largely a result of volcanic activity.

After a disastrous smog in the London area in 1952, a series of Clean Air Acts has prohibited the burning of coal in open fires over much of Great Britain. Other countries have taken similar steps, so that apart from the specialized automobile smog of the Los Angeles area, urban smogs have largely disappeared from much of the temperate world. This success has, unfortunately, not solved the SO_2 pollution problem. Houses and factories still need energy, and this is provided by burning oil and coal in power plants. Urban pollution has been avoided partly by cleaning the

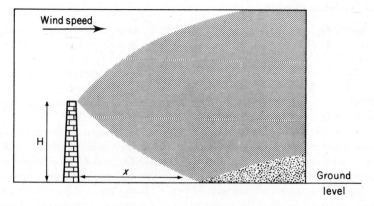

Fig. 3–1 Spread of smoke from a factory chimney. The horizontal distance x before the smoke reaches the ground depends on the windspeed, on the square of the height of the smokestack (H) and on the temperature of the flue gases. Downwind from x, the smoke concentration will be intensified (shown by the heavily shaded area), because of reflection from the ground surface.

smoke, and partly by emitting it into the upper atmosphere. The principle is shown in Fig. 3–1. Stacks 200, or even 300 m, high have been built, from which much of the gases enters the upper atmosphere where there is perpetual turbulence, so they are swept away and diluted.

There is a catch phrase that says 'dilution is the solution to pollution'; this is not true indefinitely. Europe produces about 60 million tonnes of SO_2 a year from burning fossil fuels, although oil refineries do make strenuous efforts to remove sulphur, and organic sulphides, during processing. Much of this pollution reaches southern Norway and Sweden, after the SO_2 has been oxidized to SO_3 in the atmosphere. The resulting H_2SO_4 is precipitated with snow and rain. As the prevailing winds are westerly, Britain is regarded as the villain of the piece, but this is not entirely true, as Fig. 3–2 shows. The soil in southern Norway is thin, on a granite bedrock, so that H_2SO_4 accumulates in lakes and rivers to a concentration which kills off all fish ($pH < 4.5$). Serious damage to growing conifers has been reported from Sweden. Pollution of a similar kind occurs in eastern Canada.

Sulphur dioxide pollution is not likely to exterminate mankind, but it is highly unpleasant, and it emphasizes that pollution is often a trans-continental affair, whose results cannot simply be exported and forgotten.

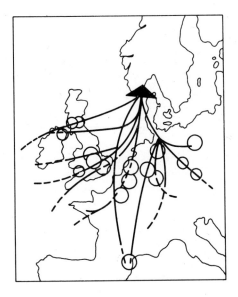

Fig. 3–2 Accumulation of SO_3 over southern Norway. The arrows show the prevailing winds on 12 successive days of January 1974. The open circles represent centres of population or industry in Western Europe.

3.3 Fluoride

Fluoride is a local pollutant, whose effects are serious in the neighbourhood of brickworks. At the high temperature of firing, any fluoroapatite in the clay is decomposed to produce eventually the volatile HF. Fluorine itself is too reactive to remain as such in the flue gases.

The fluoride ion has a rather complicated biochemistry. It occurs naturally in plant ashes at about 0.1%, and also in low concentrations (a few ppm) in some natural waters. People habitually drinking such water suffer very little from dental decay, so fluoride is often added to other sources of drinking water to a concentration of about 1 ppm to protect the teeth of the general population. It does this by replacing some of the OH^- ions in the complex calcium phosphate mineral *hydroxyapatite*, which forms the inorganic part of both bone and the enamel of teeth. Fluoride-containing hydroxyapatite is more resistant to attack by acids formed by bacteria living on traces of foodstuffs in the mouth. There is no evidence that fluoride at this concentration accumulates in the body, or has any harmful effects.

If, however, the fluoride concentration in drinking water is as high as 20 ppm, too many OH^- ions are replaced by F^- in hydroxyapatite, and the bone then becomes very hard, leading to inflammation of the joints, a condition known as 'fluorosis'. It can occur both in animals and man. If animals graze exclusively on pastures polluted with fluoride, they may suffer so severely that they have to be destroyed. At concentrations much higher than this (1 mM, or 20 000 ppm) fluoride will inhibit several enzymes that metabolize phosphate-containing compounds, sometimes by forming a very slightly ionized magnesium fluorophosphate. *Enolase* is the most sensitive enzyme; the fact is sometimes used for preventing glucose metabolism in blood containing red cells.

3.4 Heavy metals

3.4.1 Smelter effluents

Almost all the toxic effluents from smelters are distributed by the atmosphere, but as pollutants they may be more important in soil, where they can remain for a very long time; for example, arsenate-impregnated areas around disused copper smelters can often be recognized by the absence of vegetation. Other toxic metals emitted are copper and mercury. Lichens may be used to monitor the outputs (see HAWKSWORTH and ROSE, 1976).

The most toxic metals are nickel, chromium and cadmium. Compounds of nickel, chromium and arsenate cause lung cancers. Several nickel compounds, including the volatile nickel tetracarbonyl, are carcinogenic, whereas only a few chromium dusts, related to chromate, have this effect. As is usual, tumour induction is a result only of prolonged exposure, but nothing is known of the mechanism.

Cadmium oxide 'smoke', which is often an impurity in zinc smelting, is readily absorbed by the lungs and is very toxic, but not, on present evidence, carcinogenic. The most serious symptoms have been reported from Japan, as 'itai-itai' disease. This manifests itself as osteomalacia (a thinning of the mineral substance of bone) accompanied by severe pain. Reports that cadmium accumulation in the kidney leads to dangerously high blood pressure have not been substantiated.

Cadmium toxicity may be modified by sequestration in *metallothionein*, a small protein, of which one-third of the amino acids are cysteine. It is probable that the many -SH groups form multi-dentate coordinate complexes with heavy metals (cf. p. 55). The protein was first isolated as a cadmium protein (from horse kidney), but it can contain several other metals, notably zinc, mercury and silver. Administration of any of these metals to experimental animals induces metallothionein synthesis in liver, with accumulation of the metal ion in that organ, but it seems likely that the natural function of the protein is to store zinc. The metal-loaded protein later accumulates in the kidneys, leading to suggestions that it is secreted in urine. It is indeed secreted by kidney tubules, but reabsorption is so rapid, even in damaged kidneys, that it is unlikely that metallothionein is a detoxifying agent.

3.4.2   Lead pollution

The toxicity of lead has been known for a long time. Formerly, the chief sources of lead poisoning were lead-based paints, often ingested by children, and lead storage tanks and pipes carrying drinking water.

Lead plumbing has been used at least since Roman times. It has been seriously suggested (though not often accepted) that the decline of the Roman Empire was due to the infertility of Roman matrons exposed to lead poisoning from their plumbing. The extent to which Pb dissolves depends on the hardness of the water; several ppm can be found in water that is soft and slightly acid, especially if it contains natural chelating agents (humic acids) derived from peat. Recent studies from the Glasgow area have suggested that children of mothers living in soft-water areas are more likely to suffer from mental retardation. Although lead is very poorly absorbed from the gut, it is a cumulative poison, and can accumulate in bone (with exchange into plasma) over many years.

The situation has changed completely since the intensive use of the petrol engine, beginning about 1910, when the start of a rise in the Pb content of circumpolar snow can be demonstrated. The four-stroke petrol engine makes very strict demands on its fuel – it must vaporize readily when drawn into the cylinder, but burn relatively slowly when ignited. Explosion ('knocking') leads to lowered efficiency, overheating of the cylinder, and mechanical stress (see BAILEY *et al.*, 1979)

Straight-chain hydrocarbons of the correct volatility, such as *n*-heptane (C_7H_{16}), are very poor fuels, largely because the initial reaction with O_2 produces 'free radicals' (molecular fragments containing an unbonded electron). If more than one free radical is produced at each step, the combustion rate increases autocatalytically. Branched-chain hydrocarbons such as iso-octane burn much more slowly, because multiple free radical formation stops at the branch points. Unfortunately, even the most sophisticated refining technology does not produce enough branched-chain hydrocarbons of the right volatility for petrol engines. (Diesel engines do not suffer from this problem.)

The solution has been to slow down the rate of combustion by using substances known as chain-breakers or scavengers ('anti-knock agents'), the most successful of which are tetraethyl- and tetramethyl-lead. As much as 0.8 ml of such compounds have been added to each litre of petrol, equivalent to about 2 g Pb litre^{-1}; the present European limit is about 0.5 g litre^{-1}. The total amounts of lead used are staggering – 300 000 tonnes a year in the U.S.A., 50 000 tonnes a year in the U.K.

Lead tetra-alkyls are themselves extremely poisonous, volatile compounds, affecting the central nervous system, but it is the inorganic lead in the products of combustion which gives cause for concern. Fine particles of lead metal or lead halides are emitted and taken up into the lungs. Lead appears to be absorbed very much more efficiently into the bloodstream from the lungs than from the intestine.

Most of the lead particles fall to earth within a relatively short distance of the roadway. Both airborne particles and the lead-contaminated foliage are much reduced after a distance of about 150 m from the road. Airborne lead pollution is therefore essentially an urban (or motorway) problem. It has been very difficult to decide how serious it is because of the imprecision of the clinical symptoms to be expected in people suffering from sub-acute lead poisoning.

The symptoms of severe lead poisoning are well known. The most characteristic are a painful intestinal colic (spasm), loss of function in peripheral nerves leading to tremors and paralysis, failure of kidney function, and convulsions, which may be fatal. If poisoning is chronic, there may be anaemia. All but the most severe cases can be cured by prolonged treatment with a chelating agent such as EDTA. It is, however, worrying that some children who have contracted acute lead poisoning, and have been treated until their clinical symptoms have disappeared, have much later shown evidence of emotional difficulties and mental retardation. Evidence connecting the latter with high levels of lead in drinking water has already been mentioned, and the blood lead levels in children kept in institutions because of emotional instability are often high. All this suggests that exposure to lead may have a long-term effect, to which children are particularly susceptible, persisting after clinical symptoms have been relieved.

Unfortunately the milder forms of the symptoms described above – fatigue, depression, digestive upsets and mental retardation – are hard to quantify, as many of us often suffer from the first three. How does one tell whether a city-dweller is suffering from sub-acute poisoning which may have a lasting, if delayed, effect? Two approaches have been used. One is to measure the lead concentration in the blood of city-dwellers, and to compare the results with those found in clinically-defined lead poisoning. Figure 3–3 shows that, at least in the late 1960s, the lead concentration in the blood of city children was disturbingly high, although probably never equal to the threshold value for the appearance of clinical symptoms. The lower values indicated on the graph, the 'toxic thresholds', are disputed because they correspond to the mild symptoms which are so hard to evaluate. It is also noteworthy that the blood lead concentration in the *average* inhabitant of Los Angeles is the same as that of natives of New Guinea. The concentration of Pb in the milk teeth of children is more accurately diagnostic; it is distinctly high in urban children.

The other approach is to see if there is any biochemical lesion that can be used as an indicator. Pb^{2+} inhibits many enzymes, but especially those connected with the synthesis of haem, shown diagrammatically in Fig. 3–4. It is well-established that lead poisoning leads to a raised excretion of δ-aminolaevulinic acid (ALA) and porphobilinogen, and that free porphyrin accumulates in red blood cells. The excretion of ALA is the most sensitive indicator of lead poisoning, but it does not correlate at all well with the symptoms that cause most concern – those due to malfunc-

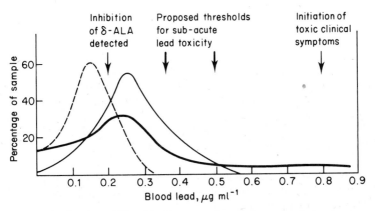

Fig. 3–3   Distribution of blood lead content from urban populations. – – – – – represents adult males living in the suburbs, ———— represents adult males living in the city centre, (both samples taken in Philadelphia, 1961–2), ———— represents normal children living in the Manchester area (1966–7). The results for the latter are not normally distributed, and the graph shows signs that it could be decomposed into two curves, for inner-city and suburban inhabitants, corresponding to those for the USA males. The outer arrows represent respectively the concentrations at which inhibition of δ-aminolaevulinic acid can be detected (left), and the threshold for the initiation of toxic clinical symptoms (right). The central arrows indicate blood lead concentrations which have been proposed as thresholds for sub-acute lead toxicity.

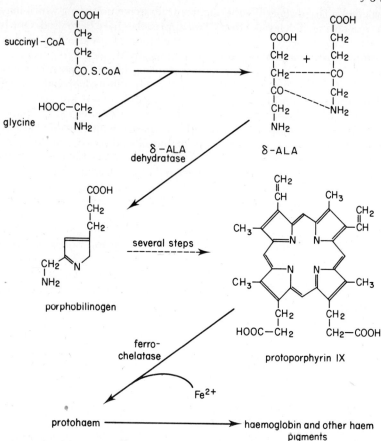

Fig. 3–4 Stages in the synthesis of haem. δ-ALA is an abbreviation for δ-aminolaevulinic acid, which can be detected in urine if the enzyme δ-ALA dehydratase is inhibited by lead.

tion of the central nervous system. Anaemia, as a result of lead poisoning, occurs rather late, whereas the nervous symptoms appear early, and are almost certainly not due to anoxia.

3·5 Aryl hydrocarbons

The word 'aryl' means having an aromatic hydrocarbon ring system. The simplest such hydrocarbon is *benzene* (C_6H_6)

or

Benzene itself is highly toxic, causing chromosome damage, particularly in bone marrow. Its nearest relative, toluene, is not harmful in this way.

Many hydrocarbons possessing several benzene rings fused together are carcinogens. They form only a small proportion of all organic compounds known to be carcinogenic, many of which are dyes, but they are of interest as some are known aerial pollutants. The question of what chemical structures are required to induce tumour formation, and the underlying biological mechanism, is too complicated to be discussed, even briefly, in this book. However, there are implications that are important when considering certain widespread atmospheric pollutants.

It has long been known that many carcinogenic hydrocarbons, especially those with a non-linear structure (Fig. 3–5, A and B), have a region of special electron density in their molecules called the 'K region', and it was thought that this was connected in some way with their carcinogenicity. More recent research (see SMITH et al., 1978) has suggested that a region between the exposed benzene ring and the rest of the molecule – the 'bay region' – may be more important. This is because it has been found that the exposed ring may be attacked *twice* by a mixed-function oxidase to give a *diol epoxide* (Fig. 3–5A(c)). Note that the ring which was attacked has now lost all its aromatic character; the second epoxide to be formed rearranges to give a very reactive *carbonium ion* (indicated by the + sign) right in the bay region. The ability of this derivative (labelled the ultimate carcinogen in Fig. 3–5A) to react with DNA (or perhaps with a cell surface component) lasts only for a few minutes. After this, it must be presumed to have rearranged to form a harmless derivative, or to have reacted with an intracellular scavenger, such as glutathione.

The correlation between the possession of a bay region and carcinogenicity *in vivo* is not perfect, but it is highly suggestive. Moreover, the evidence that a short-lived reactive intermediate is involved makes sense of empirical biological tests. In the Ames test, a suspected mutagen is incubated with a bacterial suspension. An increase in the number of mutants after incubation is a positive response. Pre-incubation of the suspect compound with a preparation of liver microsomes and $NADPH_2$ may greatly increase the mutant count at the end of the test, but the greater the interval betweeen the two incubations, the less is the effect. This is clear evidence that the liver microsomal mixed-function oxidases have the capability to produce short-lived mutagens.

Aryl hydrocarbons are not at all volatile. Traces are often formed in over-heating of animal fats or cooking oils, and they may be implicated in the development of stomach cancers. However, their chief source is undoubtedly smoke, perhaps cigarette smoke, but certainly industrial flue gases. The motor car is only a minor source; it has been estimated that car exhausts account for 50 tonnes $year^{-1}$ of benzpyrene, while coal-burning produces 4000 tonnes $year^{-1}$ on a global scale. This is substantiated by measurements showing the concentration of benzpyrene to be as much as

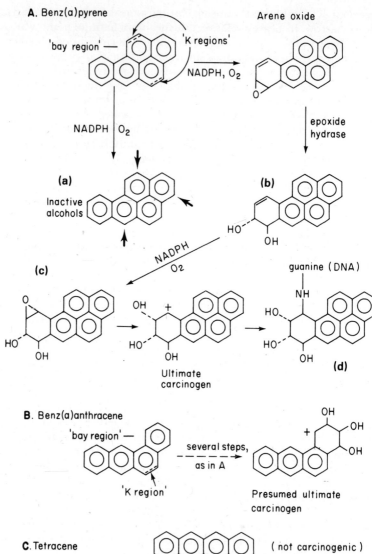

A. Benz(a)pyrene

Arene oxide

'bay region' — 'K regions'

NADPH, O_2

NADPH | O_2

epoxide hydrase

(a)

Inactive alcohols

(b)

HO

OH

(c)

NADPH O_2

guanine (DNA)

NH

HO

O

HO

OH

OH

+

HO

OH

HO

HO

OH

Ultimate carcinogen

(d)

B. Benz(a)anthracene

'bay region' —

several steps, as in A

'K region'

OH

OH

+

OH

Presumed ultimate carcinogen

C. Tetracene

(not carcinogenic)

Fig. 3–5 **A.** Benz(a)pyrene is usually attacked at the points shown by the heavy arrows in (a), to form one or more inactive phenols. A different mixed-function oxidase may attack it on the exposed benzene ring (see text, p. 35), to give first a diol (b). This is attacked a second time to give a diol epoxide (c), which rearranges to form a reactive carbonium ion. This readily reacts with DNA, and other materials, to give products such as (d). **B.** A similar chain of events is thought to occur for the more weakly carcinogenic benz(a)anthracene. **C.** The linear tetracene does not form a diol epoxide.

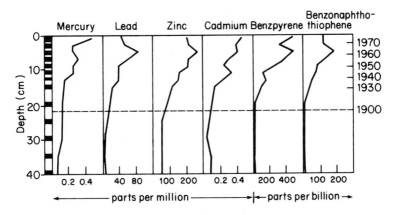

Fig. 3–6 Deposition of toxic materials in the sediments on the floor of Lake Constance, in the Swiss Alps. Note that the big peaks for lead, zinc and benzpyrene in the middle 1960s coincide with that for benzonaphthothiophene (BNT), which is a product of coal burning, but is not emitted by petroleum-powered vehicles. The deposition of mercury has been much more constant over a long time period than that of other pollutants. (From MÜLLER *et al.* (1977). *Naturwissenschaften*, **64**, 427.)

0.1 mg l^{-1} in water around industrial plants on the Caspian Sea. The correlation with coal- rather than oil-burning is demonstrated by Fig. 3–6. The rise in concentration of benzpyrene in the bottom mud is exactly paralleled by the rise in benzonaphthothiophene, a compound formed by combustion of coal, but not of oil. A rise in the deposition of heavy metals, exactly similar to this graph, is found in the marine muds off the eastern shore of the U.S.A. (TUREKIAN, 1975).

Very long exposure to carcinogens is usually necessary before a tumour is initiated. Even if all chemical carcinogens are mutagens, which is by no means certain, the efficient repair mechanisms which cells possess for excising and replacing damaged DNA must be taken into account. If every point mutation led automatically to tumour initiation, man would hardly have survived the multiplicity of challenges from noxious chemicals in his natural, and his self-made, environment. Nevertheless, one cannot afford to reject any of these chemical challenges as unimportant. It can be shown that feeding human volunteers on barbecued steak, a method of cooking known to lead to the formation of benzpyrene and similar substances on the charred outer surface, results in a most dramatic increase in the rate of destruction of the pain-killing drug phenacetin, if this is subsequently given to the volunteers. This is because the benzpyrene in the steak induces the formation of aryl hydrocarbon oxidases (p. 25) in the liver, the organ mainly responsible for drug metabolism. Thus apparently trivial events may produce far-reaching consequences.

4 Soil Pollution

This chapter deals almost exclusively with pollution caused by pesticides. Although almost all pesticides are sprayed in air, and the chemicals used, or their products, often find their way into water, it is the retention of pesticides in *soil* that is the major pollution problem.

To most people, 'pesticide' almost always means 'insecticide', but if we define a pest as 'a plant or animal living where man does not want it to live' (MELLANBY, 1967), the term has a much broader range – herbicides, fungicides, nematicides, molluscicides and rodenticides are also extremely important. A useful brief discussion, with examples of each class, is given in GUNN and STEVENS (1976).

4.1 Herbicides

Herbicides account for 40% of world pesticide production, and almost 66% of pesticide usage in highly-developed countries. They may be divided into three main classes: uncouplers, agents acting on chloroplasts and auxin simulators.

4.1.1 Uncouplers

These are substances that uncouple phosphorylation from the transport of electrons down the respiratory chain, and consequently reduce, by up to 80%, the rate of generation of ATP by the cell. This is quickly fatal, not only to plants, but to all forms of animal life. However, unless uncouplers are applied carelessly, or in excess, they can be used without danger. Many of them are derivatives of phenol, such as *dinitro-orthocresol* (DNOC)

which act by virtue of their ability to pass through the mitochondrial membrane with the phenolic -OH un-ionized, so transporting a proton and collapsing the mitochondrial proton pump.

Trialkyl tins are widely used as fungicides and molluscicides (e.g. in irrigation canals), often in a rather specialized way, but it is not known how they work. A precursor is incorporated into a paint or other substance from which the active alkyl tin is slowly released over a long period, thus

$$(C_4H_9)_3Sn-O-Sn(C_4H_9)_3 \longrightarrow Sn(C_4H_9)_3^+$$

Organotin compounds are a slightly worrying, although perhaps minor, pollution hazard. A wide range of them are used as polymerization catalysts, and also plasticizers, in the plastics industry, but it appears that only the trialkyl tins are biologically hazardous.

4.1.2 Agents acting on chloroplasts

Two examples may be given. *Simazine* inhibits the Hill reaction (the reduction of NADP and concomitant release of O_2) which is one of the vital reaction sequences in photosynthesis.

simazine

It is completely harmless to animals, being applied to soil, and absorbed by plants through their roots.

Paraquat has the ability to pick up a single electron from ferredoxin, a component of photosystem I. The electron is transferred to O_2, producing the extremely reactive and toxic superoxide ion O_2^-.

paraquat

The major effect is probably the peroxidation of the lipid component of cell membranes. It is very poisonous if swallowed, but is normally applied to foliage, and is rapidly destroyed by soil bacteria.

4.1.3 Auxin simulators

Compounds of this class mimic the action of natural plant hormones. When used as herbicides they lead to exaggerated, distorted growth, which eventually kills the plant. Monocotyledons are much less affected, so that these herbicides can be used for suppressing weeds in cereal crops. An example of an auxin simulator is *2,4,5-trichloro-phenoxyacetic acid*

(2,4,5-T, shown below). A natural auxin, indole acetic acid (IAA), is illustrated for comparison.

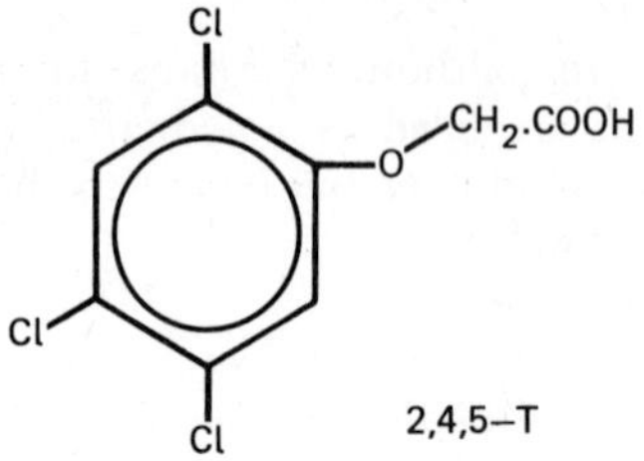

2,4,5–T 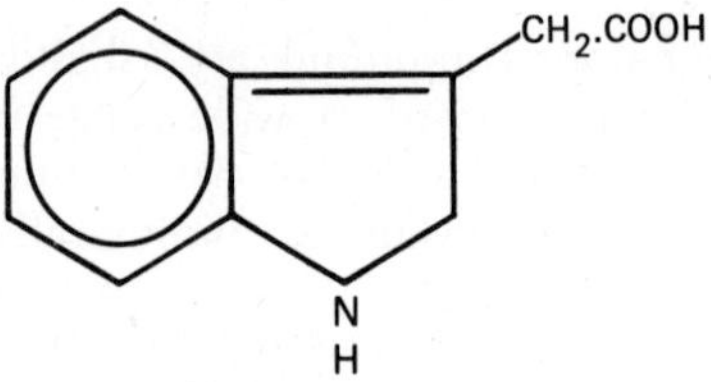   indole acetic acid

Several other phenoxyacetic acid derivatives are in use.

There are, of course, many other herbicides, many of them selective towards monocotyledons. In general, they are either not very persistent, or their action is so oriented toward plant physiology that they are not harmful to animals. The pollution hazard from herbicides is surprisingly small, considering their very intensive usage. One exception must be mentioned, a toxic by-product of the manufacture of 2,4,5-T, called *dioxin*.

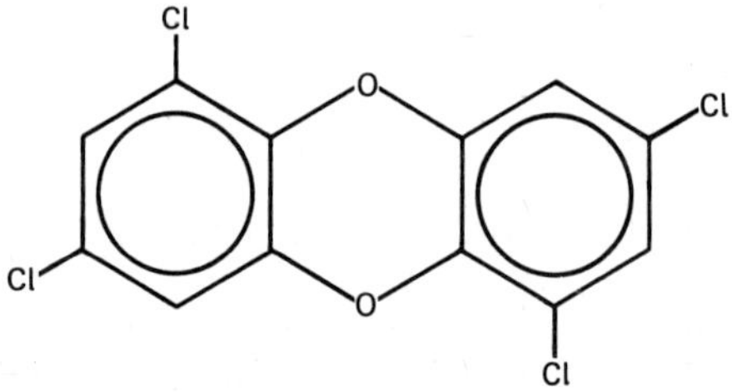

dioxin

Like other chlorinated hydrocarbons to be discussed later, dioxin has great persistence in soil and water. Large-scale release has so far been mostly accidental, as in an explosion in a factory at Seveso, northern Italy, in 1976. However, during the Vietnam war the American forces used defoliants, including 2,4,5-T, on a very large scale in order to reduce the forest cover available to their enemies. There seems little doubt that some of the herbicide was not sufficiently freed from dioxin. In very large quantities, dioxin will kill small animals, but the major effects on human beings are skin lesions, including a very unsightly and long-persisting acne, due to blocking of the ducts of the sebaceous glands. In tests on animals, dioxin has been shown to be teratogenic; in view of its persistence, it would be very alarming if this were to be confirmed for humans. Dioxin is also a very powerful inducer of liver mixed-function oxidase activity.

4.2 Insecticides

Almost all the insecticides that have been developed in the last 40 years act on the central nervous system, most of them by blocking the function of the enzyme, *acetylcholine esterase*. Two exceptions are *rotenone*, which blocks transfer of hydrogen atoms from $NADH_2$ down the electron transport chain, and the chemosterilants, such as *apholate*.

Cholinesterase catalyses hydrolysis of the ester acetylcholine, a chemical transmitter of nervous impulses across synaptic junctions. After an impulse has passed, the transmitter should be destroyed, or the efferent nerve will be kept in a state of continuous excitation. The destruction may be represented by the reaction

$$CH_3.COO.CH_2.CH_2.\overset{+}{N}(CH_3)_3 \quad + \quad \boxed{CE} \longrightarrow$$
$$\text{acetylcholine} \qquad\qquad\qquad \text{(enzyme)}$$

$$HO.CH_2.CH_2.\overset{+}{N}(CH_3)_3 \quad + \quad \boxed{CE}.OOC.CH_3$$
$$\text{choline}$$

$$+ H_2O \longrightarrow \boxed{CE} + HOOC.CH_3$$
$$\text{acetic acid}$$

The active site of the enzyme contains an -OH group, attached to the amino acid serine, which temporarily accepts the acetyl residue.

Since acetylcholine is also an important synaptic transmitter in higher animals, one might expect that the insecticides would be just as toxic to man as they are to insects. The first organophosphorus insecticides were indeed developed as nerve gases for use in warfare. The organochlorine insecticides are, however, very much less toxic to man. The selective toxicity of cholinesterase inhibitors is probably achieved in two ways. One is the rapid inactivation of the chemicals by vertebrates, but not by insects; or conversely, insects (but not vertebrates) convert inactive precursors to toxic products. The other way depends on differences in membrane structure of the synapses of insects and warm-blooded animals; it is not known precisely what these differences are. There is a strong pressure of natural selection in insect populations exposed to insecticides, which leads to the development of resistance, often by metabolizing the compounds in new ways. The history of insecticide chemistry is, therefore, of a continuous struggle to keep 'one jump ahead of the insect'. The latter term, incidentally, is often extended to include the mites and ticks (Acarina), and sometimes even the eelworms (Nematodes).

4.2.1   *Organochlorine insecticides*

Plant products, such as the pyrethrins, which kill insects have been known since the eighteenth century, but they are quickly destroyed on exposure to air. The unique quality of DDT (discovered in 1940) was its persistence, combined with insolubility in water. It remains active for up to seven years. In the Second World War DDT powder was dusted on the skin to prevent an epidemic of louse-borne typhus (a disease of high mortality) in Italy, and was used throughout the tropics for combating mosquito-borne malaria.

Strictly speaking, DDT is toxic to man, especially in oily solution, but there is no record of any fatality, and even adverse symptoms are rare. There are two reasons for this; one is that it is ineffectual in disrupting the synaptic membrane of warm-blooded animals (but not of fish); the second is that vertebrates possess an enzyme that removes chlorine from the molecule. The product, DDE, is not a cholinesterase inhibitor.

After the war DDT was introduced into farming, particularly for pest-prone crops with a high cash yield (almost 25% of all insecticides used in the world today are for the protection of cotton (ADAM, 1976)). This was followed by the commercial release of many other chlorinated hydrocarbon insecticides, a few of which are shown in Fig. 4–1. A new era of pest control was announced; it was confidently predicted that malaria and other diseases transmitted by insect vectors would be eliminated. A quarter of a century later, the use of this type of insecticide is banned in the U.S.A. and many other Western countries. Malaria has not been eradicated. Some biologists have called DDT 'the greatest disaster ever inflicted on the environment' (BONEY, 1975).

What went wrong? The answers are complex, and it would be fair to say that agricultural scientists still take a view of their own on the question. Briefly, however, disenchantment had three causes.

(1) Overuse of insecticides

Especially in the early years, there was gross over-use of the new insecticides. CARSON (1970), whose book is a salutary history of the worst period, tells of a farmer spraying toxaphene at a density of 450 lb to the acre (50 g m^{-2}). Toxaphene, heptachlor and endrin are known to have killed people, particularly in Nicaragua. Small animals were much more at risk; endrin has been used as a rodenticide, and chlordane is still on sale as a worm-killer. When used at such high concentrations, the chlorinated hydrocarbons were leached into rivers, where they poisoned fish. As the compounds are very lipid-soluble and accumulate in body fat, they move rapidly up the food chain, with disastrous results to birds. Aldrin and its associated compounds were for some years used for treating seeds, and for freeing soil of nematodes, so that almost all kinds of insects and also earthworms contained high concentrations of pesticides at the peak period of usage. There were many instances, particularly in the U.S.A., of direct death of birds. More generally, there was a marked decrease in bird populations, owing to a decrease in fertility and in hatchability of eggs, perhaps because one of the isomers of DDT has distinct oestrogenic activity, but induction of liver mixed-function oxidase activity may also have been a contributory cause.

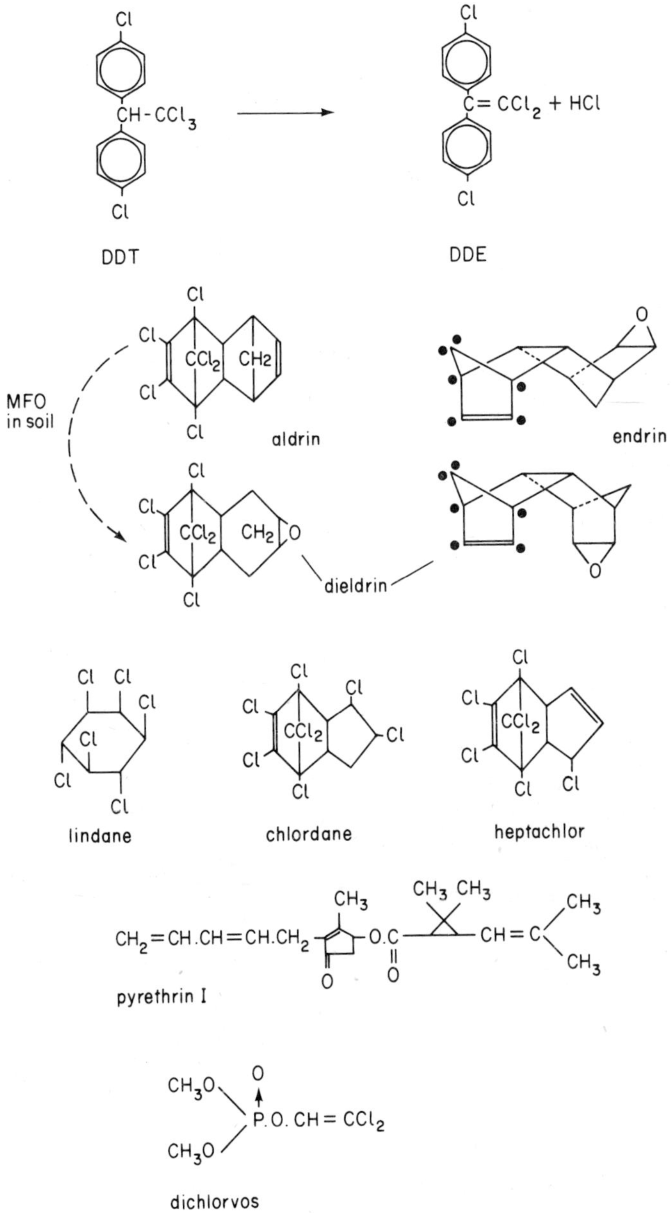

Fig. 4–1 Structures of organochlorine and other insecticides. Pyrethrin and dichlorvos are not organochlorine insecticides (although the latter contains two atoms of Cl). Dieldrin and endrin are isomers, differing only in three-dimensional structures.

(2) The stability of the organochlorines

This stability became a handicap. Chlordane, aldrin and similar compounds have been found to persist in soil for 10–12 years. Since world production of organochlorine compounds, at its peak in the early 1970s, was little less than 1 000 000 tonnes year^{-1}, the concentration of these compounds in the environment increased much faster than their rate of destruction. Even now all parts of the globe, aqueous and terrestrial, contain a detectable amount of organochlorine compounds.

(3) Development of resistance

Biologically, the most serious phenomenon was the development of resistance. Houseflies resistant to DDT appeared in Italy only five years after its first use. They had acquired the dechlorinating enzyme shown on p. 43. Twenty years later almost 100 species of medically important insects and over 200 agricultural pests had developed resistance to insecticides. Resistance develops also to the newer insecticides, but the sharp rise in the curve after 1955 (shown in Fig. 4–2) was due to the rapid development of resistance (by a means still unknown) to dieldrin and benzene hexachloride (lindane).

Insects rarely, if ever, lose their resistance once acquired, and they become resistant to a group of insecticides, rather than to a single compound. This is not a defect peculiar to organochlorine insecticides, but

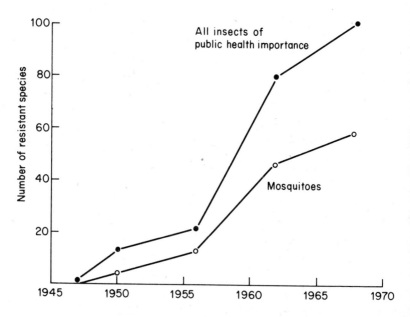

Fig. 4-2   The number of arthropods of public health importance which have developed resistance to insecticides.

resistance usually develops only after prolonged exposure. It is unfortunate that the long life of organochlorines in water and soil means that persistent exposure is virtually inevitable. Heavy spraying for agriculture has sometimes led to the acquisition of resistance by local mosquitoes.

Hopes that malaria will be eradicated by mosquito control have been tacitly abandoned. Indeed, world health authorities are now worried about a recrudescence of the disease. In absolute terms the success, at the time of its maximum, was huge. About 2×10^9 people live in potentially malarial areas. Twenty-five years ago the annual incidence of the disease was 3×10^8 cases, with 3×10^6 deaths. Until recently, the incidence had been reduced to 5×10^7 cases, with 1×10^6 deaths. Resistance to residual insecticides is only part of the reason why the number of deaths is not likely soon to fall below a million a year, but it is difficult to believe that eradication of the mosquito would ever have outpaced the development of resistance, particularly with endrin.

There are other biochemical factors that cause concern. DDT is a potent inducer of hepatic mixed-function oxidases (p. 25), aldrin and the polychlorinated biphenyls (see below) even more so. It does not necessarily follow that organochlorines are carcinogens or co-carcinogens; research in experimental animals has shown that DDT reduces the incidence of mammary tumours induced by one powerful carcinogen, presumably by the same kind of mechanism that was outlined for benzpyrene in Chapter 3 (Fig. 3–5A (a)). However, it is worrying that DDT is powerful enough to have any effect at all, since we cannot be sure that the effects of pesticides will always be favourable; we cannot even specify all the xenobiotics (foreign chemicals) in the environment.

Synergists, whose mode of action is discussed in section 4.3, have been used in conjunction with organochlorine insecticides.

4.2.2 Polyhalogenated hydrocarbons

The so-called *polychlorinated biphenyls* (PCB) are widely used as plasticizers, to render objects made of plastics less brittle, and as high-performance electrical insulators. *Polybrominated biphenyls* (PBB) are also used as fire-retardant compounds. A major component of the PBBs is shown below:

They have no insecticidal activity, but possess many of the unwelcome biological properties of the organochlorines, particularly long persistence, lipid solubility leading to concentration in food chains, and toxic effects on animals. Many deaths of sea-birds on British coasts have been attributed for several years to uncontrolled release of PCBs. An accidental but widespread distribution in 1975 throughout Michigan in the U.S.A., when a consignment of PBB was accidentally mixed into animal feedstuff, put the toxicity beyond doubt. Many farm animals were seriously affected, chiefly by malfunction of the intestinal system and brain; in some areas whole herds were destroyed and then burnt. There was widespread concern for the human population, because the PBBs were secreted into milk, and could be shown to be concentrated in human adipose tissue. Apart from mild neurological effects, there has been no immediate human toxicity, but it is now clear that polyhalogenated biphenyls are not harmless substances. They are very potent inducers of hepatic mixed-function oxidases.

4.2.3   Organophosphorus insecticides

These compounds work in the same fashion on *all* nervous systems. They inactivate cholinesterase, and so cause a lethal accumulation of acetylcholine. Differential toxicity is achieved by synthesizing variants of the basic formula which are either converted to a lethal form by the insect, or are rapidly detoxified by higher animals. The underlying mechanism of inactivation is discussed below.

The inhibitors are derivatives of phosphoric acid which has been activated in such a way that it can form an ester with the -OH group at the active centre of cholinesterase (p. 41). Derivatives that are useful as insecticides contain lipophilic groups, so that the compound can penetrate the lipid-rich membrane of the synapse to the site of the enzyme. For example with the nerve gas di-isopropyl fluorophosphate (DIPP) the two isopropyl groups confer lipid solubility, and the F atom

acts as a 'leaving group' from which the phosphate derivative is transferred to the active site of the enzyme. The resultant ester is not easily hydrolysed, so the cholinesterase is permanently inhibited. The fluorophosphates are too toxic for use as insecticides. Most of the organophosphorus compounds presently in use are tri-esters, although an exception is *tetra-ethyl pyrophosphate* (TEPP), which is very toxic to humans, but which can be used as a fumigant.

If the non-ionizing oxygen atom in $(HO)_3 P \rightarrow O$ is replaced by sulphur, thus $(HO)_3 P \rightarrow S$, the derivatives formed are much less reactive, since S is much less electronegative than O. This difference forms the basis of much selective toxicity, because mixed-function oxidases can replace S by O in a so-called 'desulphuration' reaction. Two examples are shown below.

Parathion is desulphurated to the active compound paraoxon, which then reacts with the HO- group of cholinesterase, leaving

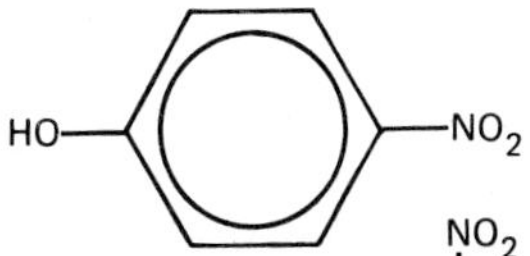

behind. Unfortunately human livers remove

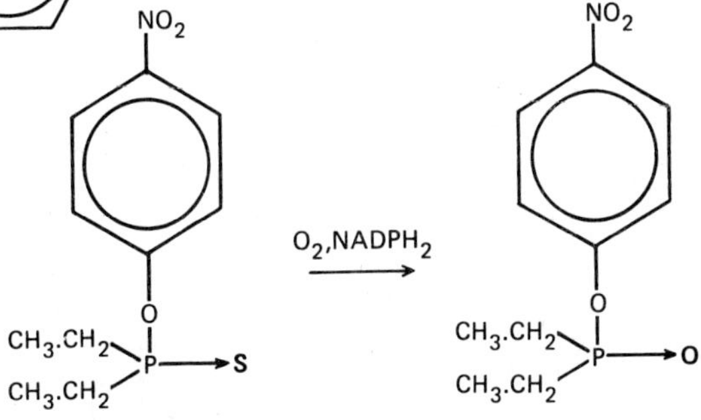

parathion   paraoxon

the S atom just as rapidly as insect fat bodies, and parathion is therefore very dangerous to humans, although still widely used.

Malathion is metabolized in a different way by higher animals, and is therefore much less toxic.

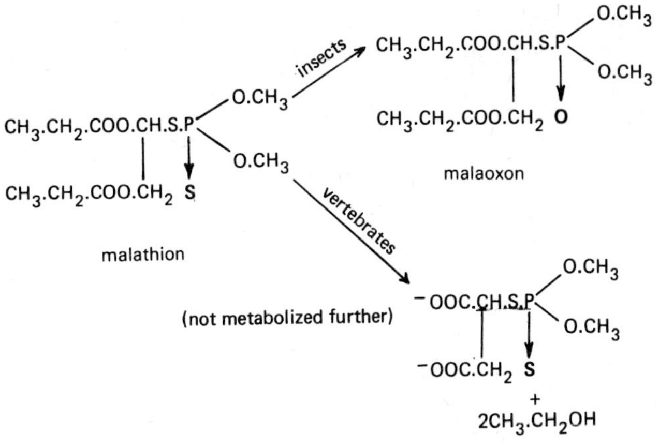

malathion

malaoxon

(not metabolized further)

Dichlorvos is also differentially metabolized, and is harmless to vertebrates. It is currently much used as a household insecticide.

Apart from the systemics, organophosphorus insecticides, and the carbamates mentioned below, are not at all persistent. This considerably reduces the pollution hazard, but makes it very difficult to use these compounds for large-scale elimination of insect vectors.

dimethoate   inactive product

Systemic insecticides are compounds which are taken up by plants, either through leaves or roots. They remain in the sap, and are absorbed by biting or sucking insects. Many systemics are organophosphorus compounds, with varying persistence in the plant and varying toxicity to vertebrates. For example, *dimethoate* is detoxified by mammals but desulphurated by insects. On the other hand, *schradan* (which is related to TEPP, p.47) and systox are highly toxic to vertebrates. There are also very useful systemic fungicides.

Systemics are mostly used for high-intensity cash crops, such as flowers, fruit or vegetables. Except for sugar-cane, they have hardly been used for field crops. Since the persistence can vary from several weeks to a few days, and the toxicity covers a wide range, the grower has to have a rather sophisticated knowledge of the product he is using. Pest attack can easily come after the latest date at which it is safe to spray before picking, and there is a temptation to let a better economic return endanger consumer safety.

4.2.4   Carbamates

These are a newer development than the organophosphorus insecticides, and are in general less toxic to higher animals. They are derivatives of carbamic acid, $HO.C{\nearrow}^{O}_{\diagdown NH_2}$, and again form stable esters with the active hydroxyl group of cholinesterase, although some carbamates may only inhibit the enzyme reversibly. One of the best-known compounds of this class is *carbaryl*.

carbaryl

4.3   Synergists

The introduction of an -OH or epoxide group into an insecticide may inactivate it. This is often the case with the carbamate insecticides, if lipid solubility is due to the presence of an aromatic ring, and with the natural or synthetic pyrethrins. The latter have a very good 'knock-down' effect, but almost no persistence, because of their very rapid inactivation. Some organochlorine compounds are also rapidly inactivated by mixed-function oxidase attack. The effectiveness of quickly-inactivated insecticides can be much improved if the mixed-function oxidase of the insect can be simultaneously knocked out. Compounds capable of doing this are known as *synergists*; they bind to a component of the oxidase system and inhibit it. The most widely used compound is *piperonyl butoxide* (below).

Synergists are intrinsically rather dangerous compounds, because they are equally as effective in man and other vertebrates as they are in insects. Moreover, they have a biphasic action. The oxidase–inhibitor complex is stable, so that a few hours after exposure to the synergist, induction of new oxidase protein begins, to replace that which has been inactivated. Thus 24 h later the mixed-function oxidase activity of the vertebrate liver may be much higher than it was originally. This could have unfortunate effects on the sensitivity of the subject to many drugs that are metabolized by hydroxylation. There is also the risk, which cannot be estimated at present, of potentiation of carcinogens. A substance very similar in composition to piperonyl butoxide, *safrole*, which occurs naturally in many plant oils, is known to be very weakly carcinogenic. It is suspected that safrole is not a carcinogen in its own right, but acts through its ability to inhibit mixed-function oxidases. The position is even further complicated by the fact that the ease with which mixed-function oxidases are induced in human liver appears to be an inherited trait, so that different people may react with very differing intensity to the same xenobiotic stimuli. Obviously one would be much happier if the widespread use of synergists in domestic environments were abandoned.

safrole

piperonyl butoxide

5 Radiation Hazards

5.1 Introduction

It is not easy to write about radioactive pollution. We are all aware that the source of pollution which is largest in scale – if not the likeliest – is nuclear war. I do not propose to discuss this at all; the issues are largely political, and the consequences would be unpredictable. I do not even wish to discuss weapons testing, which has been responsible for the most widespread distribution of radioactive nuclides to date. However, the peaceful uses of atomic energy are with us to stay, and because the term 'atomic' conditions most of us to alarm and uncertainty, it seems right to provide some information to help each of us to form his own judgment on the issues.

Biological effects depend on radiation *absorbed*; the unit is a *rad*, and 1 rad = 100 ergs g^{-1} of energy absorbed (SI unit the *gray* (Gy) = 100 rad). Different kinds of radiation have different effects – for safety purposes only a unit called a *rem* (radiation equivalent man) is used (SI unit the *sievert* (Sv) = 100 rem). No time is specified in either definition. The safe exposure of the general public is based on a 'maximum genetic dose' of 5 rems during 30 child-bearing years, sometimes miscalled the 'lifetime dose'.

The present unit of radioactivity is the *curie* (Ci), which is based on the radiation given off by 1 gram of radium. It is 3.7×10^{10} disintegrations per second. This unit is being replaced by the *becquerel* (Bq), which is equivalent to 1 disintegration per second.

5.2 Types of radiation

α-rays are charged helium nuclei ($^4He^{2+}$). They arise from the disintegration of unstable isotopes of elements having atomic weights > 150. α-rays travel slowly and lose energy rapidly by collisions, which gives them a short range in air (*c.* 5 cm), but a high biological effectiveness (1 rem ≈ 0.05 rad).

β-rays are electrons (or positrons) which have very little mass, but which can attain speeds almost as great as that of light. β-emitters can be either light or heavy nuclei; tritium (^3H) is a β-emitter. The range of a fairly energetic (1 MeV) β-particle in air is about 3 m (1 rem ≈ 1 rad).

γ-rays are not particles, but quanta of electromagnetic radiation, with much shorter wavelengths, and hence much more energy, than rays of visible light. They have much in common with X-rays. γ-rays can be emitted both on disintegration, or when a nucleus captures another particle, such as a proton or neutron. They have a very long range in air, 4 m or more (1 rem ≈ 1 rad).

Neutrons (*n*) are particles with unit mass and no charge, which are only liberated when susceptible elements are bombarded with α- or γ-rays, usually from the heaviest elements (the actinides). They only react with other elements by direct collision, not ionization, and so they have a very long range in air (1 rem ≈ 0.2 rad).

Neutrons and γ-rays will pass through several centimetres of lead, but α- and β-rays can be stopped by quite thin sheets of light materials, such as aluminium or perspex.

5.3   Biological damage caused by radiation

This is the only thing of interest from our point of view. α-, β-, and γ-rays cause damage by raising atoms, which they approach or collide with, to a higher energy level; often an orbital electron is lost altogether (ionization). The molecule to which the excited atom is bound becomes temporarily much more reactive. Most damage is caused if a chain reaction is set up. Chemicals such as *cysteamine* ($HS.CH_2.CH_2.NH_2$) are very effective at preventing chain reactions, and can protect substantially against radiation damage, but only if taken before irradiation starts.

Because of their very limited range, α- and β-rays are really only effective if they are inside the body, whereas γ-rays and neutrons are very dangerous outside the body. For example, they are especially dangerous to the eyes, leading to cataract formation.

Neutrons mostly disrupt molecules (including water) to give a 'recoil proton' (H^+) and a reactive, negatively-charged ion, but they can also transmute elements. The elements most affected vary with the energy of the neutrons, but the reaction

$$^{31}P + n \longrightarrow {}^{28}Al + \alpha + \gamma$$
followed by $\quad {}^{28}Al \longrightarrow {}^{28}Si + \beta$

is potentially most damaging, not only because an α-particle is produced, but also because the transmutation of any P atom in DNA breaks the molecule. Other elements likely to capture neutrons *in vivo* are N, Na, Cl and Ca.

It is in general the nucleus of cells that is most sensitive to radiation damage. X- and γ-rays are used for cancer therapy, or for sterilizing male insects in one method of biological pest control. Thus the tissues most at risk are those where there is a high concentration of nuclei with frequent cell division, for example, bone marrow. Radionuclides absorbed into bone are particularly dangerous because they tend to stay there a long time, whereas atoms in other sensitive sites such as liver or intestinal epithelium are much more rapidly metabolized.

The DNA of the nucleus is very readily damaged by radiation which causes ionization. In theory, it only needs one chemical change in a single purine or pyrimidine base of a stretch of DNA for a point mutation to occur, because there will be a misreading on the next occasion when that length of DNA is transcribed. In practice, however, cells contain rather efficient repair mechanisms for replacing damaged stretches of DNA. One of the ways in which we know this is from a study of the very rare disease *Xeroderma pigmentosum*, in which there is a much increased rate of formation of skin cancers after exposure to sunlight. It has been shown that the epithelial cells of sufferers from this disease lack the normal nuclear repair mechanisms. Thus, the cell's repair mechanism has to be overwhelmed by radiation damage before a mutagenic event can be guaranteed; indeed, the effect of intense bursts of radiation may be to damage the enzymes involved in nucleic acid metabolism rather than the DNA itself. This efficient repair mechanism is the reason why radiation given for therapy or sterilization has to be so intense, and it means that low doses of radiation are not as dangerous as they might at first appear. A similar persistent exposure to a noxious agent before cell transformation is certain is also characteristic for the chemical carcinogens referred to in Chapter 4. Of course, one can never be certain that a single chemical change in a cell will not cause a mutation in one cell that later becomes cancerous. It is therefore extremely difficult to be sure that there is a threshold, either for carcinogens or for radiation, below which the body is safe from attack. For safety's sake, it is assumed that, with radiation, there is no threshold level.

5.4   Half-life

This term has a very precise meaning. Each radioactive *nuclide* (a term to be preferred to *isotope*) has a characteristic instability, which is not altered by physical treatment or by formation of chemical bonds. It is impossible to predict when an individual atom will decay, but one can say that, of a finite assemblage of atoms, half will have decayed by the end of a time interval characteristic of that nuclide (see Fig. 5–1). This time interval is the *half-life*. At the end of a second half-life, half of the unstable atoms remaining will have decayed, and so on. Half-lives vary enormously in magnitude, from seconds to billions of years.

5.5   Nuclear energy plants

A *thermal reactor* (Fig. 5–2) uses natural uranium, which is predominantly a mixture of the isotopes ^{235}U and ^{238}U. Both decay over a very long time to various other elements, and finally to lead. However, ^{235}U, especially if it is irradiated by neutrons, disintegrates into a mixture of 'fission products' together with several neutrons. In principle, the latter make the process self-sustaining, although it is usual to start with

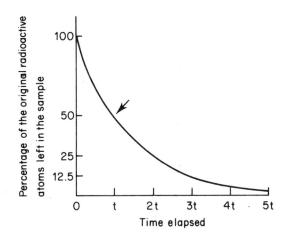

Fig. 5-1 The half-life of a radioactive isotope. The arrow indicates a single half-life.

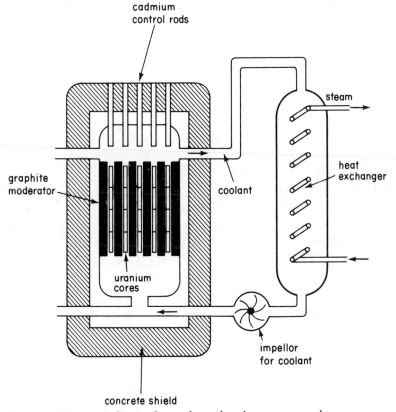

Fig. 5-2 Diagram indicating how a thermal nuclear reactor works.

uranium that has been slightly enriched in ^{235}U. Efficient neutron capture is, curiously enough, aided by slowing the neutrons down (which converts them from 'fast' to 'thermal' neutrons), using a *moderator* (graphite or heavy water, D_2O). The fission processes release energy as heat, which is used for generating electrical power in a conventional way (Fig. 5–2). One kilo of ^{235}U is equivalent to 9×10^6 kilos of coal.

After some time, typically 10–20 years, either all the ^{235}U in the reactor cores has been used up, or too many reactive fission products have accumulated. The cores are removed and 'processed', which may simply mean removing the ^{238}U, and converting the fission products into forms which can be safely stored until their radioactivity has decayed. Capture reactions will also have occurred, leading to the formation of transuranium elements, particularly neptunium and plutonium.

$$^{238}U + n \longrightarrow \epsilon + {}^{239}Np \longrightarrow \epsilon + {}^{239}Pu$$

^{239}Pu disintegrates in a similar way to ^{235}U, but much more effectively.

The *fast reactor* uses plutonium which has been extracted from used reactor cores and separated from fission products. The extraction of useful energy needs special engineering techniques, particularly with respect to the coolant. For obvious reasons, there is little information about the long-term reliability of the engineering at the commercial level.

If the plutonium is mixed with the otherwise useless ^{238}U, the neutron flux in a fast reactor will slowly convert the latter into more ^{239}Pu, so that the process ends up with more plutonium than it had at the beginning. This is the principle of the *breeder reactor*. Technologically the fast breeder reactor is a very attractive proposition, because plutonium would be produced in any case by almost any fission process. It has, however, the disadvantage that the potentially dangerous plutonium has to be processed twice.

5.6 Biological hazards of nuclear energy production

These can arise from two separate sources – those which result from the normal running of the process, and those which might arise from a nuclear accident.

5.6.1 *Normal running*

Plutonium is extremely toxic, because it is an α-emitter which can become concentrated in bone. For reasons not clearly understood, it is 4–5 times more toxic than an equivalent amount of radium, and the permissible 'body burden' is only 0.008 *micro*curies, equivalent to 0.0005 micrograms. The element is readily absorbed into the lungs as fine particles of plutonium oxide, whence it is slowly transferred to soft tissues like liver, and to bone. Metabolically speaking, it is inert, and so is only very slowly excreted. Biochemists have, however, found a way of dealing with toxic metal ions, by forming *coordination compounds*. These are compounds in which a suitable atom donates two electrons to the metal ion, forming a bond which is neither ionic nor covalent, and not readily dissociated. If several donor atoms are combined on one molecule, and the spatial geometry is favourable, the resulting multi-coordination complex or *chelate* will be very stable indeed, and can be so arranged as to be soluble in water, and thus readily excreted. The chelating agent most suitable for plutonium is *diethylene triamine pentaacetic acid* (DPTA).

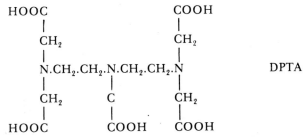

DPTA

Chelating agents can also be used to remove other heavy metals from biological systems; the best-known agent is ethylene diamine tetraacetic

Table 3   Radioactive nuclides likely to be produced by fission.

Nuclide	Major radiation	Half-life	Element with similar biochemical properties
^{3}H	β (very weak)	12.3 yr	hydrogen
^{85}Kr	β (medium)	10.6 yr	none
^{89}Sr	β (strong)	50.4 day	calcium
^{90}Sr	β (medium)	28 yr	calcium
^{91}Y	β (strong)	58 day	none
^{99}Tc	β (strong)	2.1×10^5 yr	none
^{103}Ru	β (weak) γ (weak)	40 day	none
^{129}I	β (weak) γ (very weak)	16×10^6 yr	iodine
^{131}I	β (medium) γ (weak)	8 day	iodine
^{137}Cs	β (medium) γ (medium)	30 yr	potassium
^{140}Ba	β (strong)	12.8 day	calcium (but not readily absorbed)
^{141}Ce	β (medium)	32.5 day	none
^{144}Ce	β (medium)	284 day	none
^{147}Pm	β (medium)	2.6 yr	none
^{237}Np*	α	2.2×10^6 yr	none

* Not a fission product; it arises from neutron capture by ^{235}U and ^{238}U. Traces of neptunium, like other heavy metal ions from yttrium on, may be adsorbed into bone, although they are not really biological analogues of calcium.

acid (EDTA). Care has to be taken that the chelator does not remove useful metal ions from the body, particularly Ca^{2+}, and so it is usually given as the Ca salt.

One of the weaknesses of chelators is that they do not readily penetrate cells, where the noxious ions reside. British scientists have recently developed a derivative of DPTA, called 'Puchel', which has long lipid chains attached to the molecule. This enables it to cross cell membranes easily, and it removes plutonium efficiently from soft tissues, although not from bone.

In spite of this exceedingly high toxicity, it is fair to say that plutonium is not at present regarded as a high pollution risk due to the stringent precautions taken in atomic energy plants.

Fission products   The major products are nuclides (Table 3) with roughly half the atomic weight of the fuel, i.e., *c.* 90–150. Ten of the nuclides account for over half the fuel which disappears. In addition, there are substantial amounts of tritium (^3H), and transuranium elements (actinides). One of these, ^{238}Pu, is not a fission fuel, but is used as a portable power source by virtue of the heat dissipated by absorption of the α-rays which it releases. The biological hazards of these products depend on the chemical state and reactivity of each, and on their half-lives. For example, several hundred curies of tritium are released from atomic energy plants in the U.K. every year. Krypton is an inert (noble) gas, which undergoes no chemical changes in biological systems. In the U.S.A. it is released into the atmosphere, but is more dangerous than tritium because it is heavier than air, remains mixed in the atmosphere, is inhaled passively into the lungs and dissolves, to a slight extent, in the blood.

Neither of these elements is nearly so much of a problem as the nuclides of very long half-life, for example iodine-129, because safe storage during a decay period of millions of years cannot be unequivocally guaranteed. Fission products of shorter half-life, such as ^{90}Sr, are, at the present time, simply stored until their radioactivity has decayed. This is quite difficult in terms of engineering, because large volumes are involved and some of the products are corrosive. Also, considerable heat is involved as the nuclides decay, so the storage tanks have to be continuously cooled.

5.6.2   Accidents at nuclear power plants

Most people think that if an atomic energy plant gets out of control, there will be a nuclear explosion, complete with mushroom-shaped cloud. This is very unlikely. It is true that a mass of 300 g of pure plutonium could 'go critical', but in a fast breeder reactor it is mixed with non-fissile ^{238}U in a proportion of 1:4. Moreover, a great deal of trouble is taken to separate the fuel elements with neutron-absorbing substances such as cadmium, to eliminate the risk of the core becoming critical in any circumstances.

The chief risk is that the cooling system will fail. Although the neutron flux can be slowed down considerably, which would reduce the heat given off during the fission itself, the accumulated fission products give off heat as they decay, and cannot be stopped from doing so. If all the cooling systems were out of action simultaneously for more than a few minutes, the build-up of heat might be enough to melt the array of elements and moderators in the core, so that the cooling and control systems could never again become operational. This would lead to an intense volatilization, in which the dispersal of the more volatile fission products would be the greatest danger. Nuclear engineering technology is therefore designed to prevent cooling failure, and to minimize the effects if failure does occur.

Of the fission products listed in Table 3, four would have serious effects because of their volatility and biological activity. These are ^{90}Sr, ^{129}I and ^{131}I, and ^{137}Cs. Strontium is metabolized in the same manner as calcium, and therefore accumulates in bone. Because of its relatively long half-life, it is a very worrying pollutant. Iodine is taken up by grasses, whence it is ingested by cows, and passes into their milk. The radioactive iodine taken in by humans from this source is concentrated in the thyroid gland, which then becomes irradiated. Finally, caesium is an analogue of potassium. It is taken up into the cells of soft tissues, particularly muscle. Its half-life is even longer than that of strontium, and like the latter, it is both a β- and a γ-emitter. However, bone marrow is a more sensitive tissue than muscle, and the increased incidence of leukaemia after exposure to fallout from nuclear weapons suggests that strontium is more to be feared than caesium. In adition to the fission products, there would inevitably be traces of actinide dusts, but these are so heavy that the area over which they could fall would be very limited.

Accidents at nuclear power plants have happened, both in this country and abroad (cf. MEDVEDEV, 1977). In the accident at Three Mile Island, near Pittsburgh, U.S.A., in 1979, a total failure of the cooling system, as described at the top of this page, did almost occur. The most serious accident in the U.K. happened at Windscale, Cumbria, in 1957. It is estimated that 20 000 Ci of ^{131}I, and about 700 Ci of caesium and strontium were released. It was necessary to confiscate and dump milk produced within an area of 500 km^2 for some weeks. For the sake of comparison, the present *permitted* rate of radionuclide emission from atomic energy plants in Britain is about 5000 Ci per year. About 20% of this is actually released, mostly as plutonium.

It is desirable to keep the likely exposure to radiation and radioactive nuclides as pollutants in perspective. The U.S. Radiation Protection Guide suggests that the average exposure of an individual should not be more than 170 millirems per year. It is particularly important that exposure should be low during growth and development, both before and after birth. In the light of this recommendation, it is striking that the principal exposure to radiation in Western countries comes from X-rays, used both diagnostically and for treatment. Exposure is estimated to be about 70 millirems per year, and of course often starts before birth. Exposure from all other sources of man-made radiation, both military and peaceful, is at present 4 millirems per year.

Envoi

At the end of the book the author may perhaps risk a cautious expression of opinion. How are we to reconcile two such divergent statements as the following? 'DDT is one of man's most disastrous large-scale experiments' (BONEY, 1975). '. . . the continued clamour about hazards from pesticides may well seem to be out of proportion to the dangers that accompany their use' (J. M. BARNES, in GUNN and STEVENS, 1976).

As so often with unreconcilable statements, like was not being compared with like. The first referred to the possible long-term effects of sub-lethal concentrations of insecticides, particularly in the sea, while the second referred to the likelihood of direct poisoning of human beings by insecticides.

Pesticides are indeed, with rare exceptions, much less dangerous to humans, *when properly used*, than antibiotics. Very few laymen object to antibiotics; it is doctors who have become so worried about the multiple resistances of pathogens to antibiotics that they are returning to chemical drugs, many of which were thought to have been superseded long ago. Antibiotics are nowadays often reserved for life-or-death cases. Agricultural scientists cannot afford to be less conscientious than doctors.

On the other hand, the layman should accept that the major radiation hazard at the present time is from the medical use of X-rays. When those who demonstrate against the proliferation of nuclear power stations demonstrate with equal fervour against the unsupervised exposure to X-rays of immigrants to Britain, justice will be being done.

When scientists show irritation at the layman's stubborn, and often laudable, refusal to give up his concern for the 'quality of life', they are usually concerned about world food production or world health. Plant protection agents make an enormous difference to crop yields. Moreover, animal predators must often be kept under control because they have responded to changes that man himself has made in the environment. The development of irrigation schemes, as for instance that supplied by the Aswan Dam in Egypt, leads to an enormous increase in the fresh-water snail population. Because the snails are an obligatory intermediate host in the life cycle of the fluke *Schistosoma haematobium*, the very unpleasant disease schistosomiasis has become much more frequent. It seems better to risk pollution by poisoning the snails than not to interfere. Again, we do right to ensure, by any means in our power, that rats do not spoil the crops that we grow and harvest, and keep in store. When man alters the environment on a massive scale by agriculture, he must endeavour to strike a balance with pests competing against him. It is only necessary to ensure that the weapons he uses do not themselves create a desert, as the result of greed or indifference.

Further Reading

ADAM, A. V. (1976). In *Pesticides and Human Welfare* (ed. GUNN, D. L. and STEVENS, J. C. R.). pp. 115–30. Oxford University Press.

BAILEY, R. A., *et al.* (1979). *Chemistry of the Environment.* Academic Press, New York. 575 pp.

BARNES, J. M. (1976). In *Pesticides and Human Welfare* (ed. GUNN, D. L. and STEVENS, J. C. R.). pp. 181–92. Oxford University Press.

BONEY, A. D. (1975). *Phytoplankton.* Studies in Biology no. 52. Edward Arnold, London. 124 pp.

BRODINE, V. (ed.) (1975). *Radioactive Contamination.* Harcourt Brace Jovanovich, New York. 190 pp.

BRYCE-SMITH, D. (1971). *Chemistry in Britain,* 7, 54–6.

BUSVINE, J. R. (1976). In *Pesticides and Human Welfare* (ed. GUNN, D. L. and STEVENS, J. C. R.). pp. 193–205. Oxford University Press.

CARSON, R. (1970). *Silent Spring.* Penguin Books, London.

DUGAN, P. R. (1972). *Biochemical Ecology of Water Pollution.* Plenum Press, New York. 159 pp.

Ecotoxicology and Environmental Safety (1977–8). Academic Press, New York. Vol. 1 onwards.

The assessment of Sub-lethal effects of Pollutants in the Sea (1979). *Phil. Trans. Royal Society,* Vol. 186, 397–633.

ELKINGTON, J. (1977). *New Scientist,* 73, 706–8.

GUNN, D. L. and STEVENS, J. C. R. (eds) (1976). *Pesticides and Human Welfare.* Oxford University Press. 278 pp.

HANLON, J. (1977). *New Scientist,* 75, 176–7.

HAWKSWORTH, D. L. and ROSE, F. (1976). *Lichens as Pollution Monitors.* Studies in Biology no. 66. Edward Arnold, London. 64 pp.

Health Hazards in the Human Environment (1972). WHO, Geneva.

JENSEN, S. and JERNELØV, A. (1969). *Nature, London,* 223, 753–4.

Lead Pollution in Birmingham (1974). DoE Pollution Paper 14. HMSO, London.

MEDVEDEV, Z. (1977). *New Scientist,* 74, 761–4.

MELLANBY, K. (1979). *The Biology of Pollution.* Studies in Biology no. 38, second edition. Edward Arnold, London. 72 pp.

Minor Constituents of the Atmosphere (conference report) (1974). *Tellus,* Vol. 26.

MORRIS, J. G. (1972). *A Biologist's Physical Chemistry.* Second Edition. Edward Arnold, London.

PARKE, D. V. (1974). In *Modern Trends in Toxicology,* 2. (Eds BOYLAND, E. and GOULDING, R.). Plenum Press, New York.

SCHLEGEL, H. G. (1974). *Tellus,* 26, 11–20.

SEILER, W. (1974). *Tellus,* 26, 116–35.

SMITH, I. A., BERGER, G. D., SEYBOLD, P. G. and SERVÉ, M. P. (1978). *Cancer Res.* 38, 2968–77.

Some Aspects of the Safety of Nuclear Installations in Great Britain (1977). HMSO, London. 48 pp.

TUCKER, A. (1972). *The Toxic Metals.* Friends of the Earth, San Francisco.

TUREKIAN, K. K. (1975). *The Oceans.* Second edition. Prentice-Hall, New Jersey. 149 pp.